For further volumes:
http://www.springer.com/series/7651

Cell-Cell Interactions

Methods and Protocols

Second Edition

Edited by

Troy A. Baudino

Texas A&M Health Science Center, Central Texas Veterans
Health Care System, Temple, TX, USA

Editor
Troy A. Baudino
Texas A&M Health Science Center
Central Texas Veterans Health Care System
Temple, TX, USA

ISSN 1064-3745   ISSN 1940-6029 (electronic)
ISBN 978-1-4939-6049-1   ISBN 978-1-62703-604-7 (eBook)
DOI 10.1007/978-1-62703-604-7
Springer New York Heidelberg Dordrecht London

Humana Press is a brand of Springer
Springer is part of Springer Science+Business Media (www.springer.com)

Preface

Cell–cell interactions, as well as interactions between cells and the extracellular matrix (ECM), are essential to the development and function of tissues and organs. While cell–cell interactions are generally dynamic, there are varying degrees of stability. Tight cell–cell junctions are stable, such as those in the heart, and play an essential role in the organization of the cells. Other interactions are transient in nature, such as interactions between cells of the immune system. Nevertheless, for the maintenance of proper form and function of all tissues and organs, cells must communicate with each other.

Cells can communicate with each other in multiple ways, including through chemical, mechanical, and electrical signals. Chemical signaling can occur through several different mechanisms. Autocrine signaling is when a cell secretes a chemical messenger that binds to autocrine receptors on the same cell, which in turn affects the way the cell functions. Paracrine signaling is a form of signaling in which the cell affects neighboring cells by secreting chemicals into the common intercellular space. In addition, cells can directly transfer ions or small molecules (miRNAs, small signaling proteins) from one cell to another through pores in the cell membrane called gap junctions. This is the quickest method of cell–cell communication and is found in tissues where fast, coordinated activity of cells is required, such as in the heart.

Cells can also respond to mechanical signals in the form of externally applied force or force generated by cell–cell or cell–ECM interactions. Many cell functions, such as motility, proliferation, differentiation, and survival, can be altered by changes in the stiffness of the substrate to which the cells are adhered or through the pull of other cells, even when chemical signals remain unchanged. Interestingly, mechanical deformation of cardiac fibroblasts can cause membrane depolarization leading to a concept of mechano-electrical transduction. Cell junctions, such as through connexins, are important for cellular communications in other organ systems and likely play similar roles in physical communication between fibroblasts and other cells within the myocardium. Indeed, it has been demonstrated through Cx43 that electrical coupling of myocytes and cardiac fibroblasts can occur. In addition, in vitro cell–cell interaction assays have shown that cardiac fibroblasts and myocytes communicate through the formation of tight cell–cell junctions. Moreover, ion channels also play an intriguing and important method of signaling because abnormalities in these channels can lead to tissue dysfunction. Clearly, it is a combination of the various signals (electrical, chemical, and mechanical) that allow for proper form and function of the tissue or organ.

While whole animal models provide insight into gene-specific mechanisms, these models are limited by the complexity of the whole organism. Therefore, the use of cell models to examine cell–cell interactions is critical for our understanding of how cells communicate and what genes or proteins are altered in disease states.

The aim of this volume of Methods in Molecular Biology: Cell–Cell Interactions is to provide a collection of protocols, incorporating in vivo and in vitro methods-based approaches. This book brings together many currently used assays in examining cell–cell interactions. It is my belief that this work will represent an important resource for researchers, which will be valuable not only to those already involved in the cell–cell interaction field but also to those who are new to the area. I hope that you will find cell–cell interactions instructive and useful in your studies.

Temple, TX, USA   *Troy A. Baudino, Ph.D., F.A.H.A.*

Contents

Contributors

KATHARINA AUSTEN • *Group of Molecular Mechanotransduction, Max-Planck-Institute of Biochemistry, Martinsried, Germany*

F. BARTSCH • *General, Visceral and Transplantation Surgery, Medical Center of the Johannes Gutenberg-University Mainz, Mainz, Germany*

TROY A. BAUDINO • *Texas A&M Health Science Center, Central Texas Veterans Health Care System, Temple, TX, USA*

STEPHANIE L.K. BOWERS • *Department of Medical Pharmacology and Physiology, University of Missouri School of Medicine and Dalton Cardiovascular Research Center, Columbia, MO, USA*

LISANDRA E. DE CASTRO BRÁS • *San Antonio Cardiovascular Proteomics Center, The University of Texas Health Science Center, San Antonio, TX, USA; Barshop Institute for Longevity and Aging Studies, The University of Texas Health Science Center, San Antonio, TX, USA; Division of Geriatrics, Gerontology and Palliative Medicine, Department of Medicine, The University of Texas Health Science Center, San Antonio, TX, USA*

JOSEPH BRESSLER • *Hugo Moser Laboratory at the Kennedy Krieger, Kennedy Krieger Institute, Baltimore, MD, USA; Department of Environmental Health Sciences, Center In Alternatives In Animal Testing, Bloomberg School of Public Health, Johns Hopkins University, Baltimore, MD, USA*

CHRISTOPHER S. CHEN • *Department of Bioengineering, University of Pennsylvania, Philadelphia, PA, USA*

ANNA CHROSTEK-GRASHOFF • *Group of Molecular Mechanotransduction, Max-Planck-Institute of Biochemistry, Martinsried, Germany*

KATHERINE CLARK • *Department of Environmental Health Sciences, Center In Alternatives In Animal Testing, Bloomberg School of Public Health, Johns Hopkins University, Baltimore, MD, USA*

DANIEL M. COHEN • *Department of Bioengineering, University of Pennsylvania, Philadelphia, PA, USA*

GEORGE E. DAVIS • *Department of Medical Pharmacology and Physiology, University of Missouri School of Medicine and Dalton Cardiovascular Research Center, Columbia, MO, USA; Department of Pathology and Anatomical Sciences, University of Missouri School of Medicine, Columbia, MO, USA*

MATTHEW T. DAVIS • *Department of Medical Pharmacology and Physiology, University of Missouri School of Medicine and Dalton Cardiovascular Research Center, Columbia, MO, USA*

DAVID E. DOSTAL • *Division of Molecular Cardiology, Department of Medicine, Texas A&M Health Science Center, Cardiovascular Research Institute, Temple, TX, USA*

TIMOTHY N. FEINSTEIN • *Department of Pharmacology and Chemical Biology, University of Pittsburgh, Pittsburgh, PA, USA*

HAO FENG • *Texas A&M Health Science Center, College of Medicine, Cardiovascular Research Institute, Temple, TX, USA*

DONALD M. FOSTER • *Central Texas Veterans Health Care System, Temple, TX, USA*

ANDREA FREIKAMP • *Group of Molecular Mechanotransduction, Max-Planck-Institute of Biochemistry, Martinsried, Germany*

ERIC GABISON • *Univ Paris Diderot and Fondation A. de Rothschild, Paris, France*

P. GASSMANN • *General, Visceral and Transplantation Surgery, Medical Center of the Johannes Gutenberg-University Mainz, Mainz, Germany*

FNU GERILECHAOGETU • *Texas A&M Health Science Center, College of Medicine, Cardiovascular Research Institute, Temple, TX, USA*

SHANNON GLASER • *Central Texas Veterans Health Care System, Temple, TX, USA; Digestive Disease Research Center, Texas A&M Health Science Center, College of Medicine, Temple, TX, USA*

HONEY B. GOLDEN • *Texas A&M Health Science Center, College of Medicine, Cardiovascular Research Institute, Temple, TX, USA*

CARSTEN GRASHOFF • *Group of Molecular Mechanotransduction, Max-Planck-Institute of Biochemistry, Martinsried, Germany*

WEI-HUI GUO • *Department of Biomedical Engineering, Carnegie Mellon University, Pittsburgh, PA, USA*

J. HAIER • *Molecular Biology Lab, Department of General Surgery, University Hospital of Muenster, Muenster, Germany*

KEVIN HAKALA • *Department of Biochemistry, San Antonio Cardiovascular Proteomics Center, The University of Texas Health Science Center, San Antonio, TX, USA*

JUAN C. IBLA • *Department of Pediatrics, Genetics Medicine and Integrative Systems Biology, George Washington University, Washington, DC, USA*

CLAUS JORGENSEN • *Division of Cancer Biology, The Institute of Cancer Research, London, UK*

M.L. KANG • *Molecular Biology Lab, Department of General Surgery, University Hospital of Muenster, Muenster, Germany*

FARAH KHAYATI • *Laboratoire de Pharmacologie and INSERM U940, Hôpital Saint-Louis, Paris, France*

JOSEPH KHOURY • *Department of Cellular and Molecular Biology, Exogenesis Corporation, Billerica, MA, USA*

DAE JOONG KIM • *Department of Medical Pharmacology and Physiology, University of Missouri School of Medicine and Dalton Cardiovascular Research Center, Columbia, MO, USA*

CARLEEN KLUGER • *Group of Molecular Mechanotransduction, Max-Planck-Institute of Biochemistry, Martinsried, Germany*

WILLIAM T. LEE • *The Laboratory of Immunology, New York Department of Health, The Wadsworth Center, Albany, NY, USA*

YAOJUN LI • *San Antonio Cardiovascular Proteomics Center, The University of Texas Health Science Center, San Antonio, TX, USA; Barshop Institute for Longevity and Aging Studies, The University of Texas Health Science Center, San Antonio, TX, USA; Division of Geriatrics, Gerontology and Palliative Medicine, Department of Medicine, The University of Texas Health Science Center, San Antonio, TX, USA*

MERRY L. LINDSEY • *Department of Physiology and Biophysics, University of Mississippi Medical Center, Jackson, MS, USA*

S.T. MEES • *Department of General and Visceral Surgery, University Hospital of Muenster, Muenster, Germany*

SUZANNE MENASHI • *CNRS EAC 7149, Faculté des Sciences, Université Paris-Est Créteil, Créteil, France*

CHUN-XIA MENG • *Department of Medical Pharmacology and Physiology, University of Missouri School of Medicine and Dalton Cardiovascular Research Center, Columbia, MO, USA*

KEITH MOORE • *Biomedical Engineering Program, University of South Carolina, Columbia, SC, USA*

SAMIA MOURAH • *Laboratoire de Pharmacologie and INSERM U940, Hôpital Saint-Louis, Paris, France*

DAMIR NIZAMUTDINOV • *Texas A&M Health Science Center, College of Medicine, Cardiovascular Research Institute, Temple, TX, USA*

PIETER R. NORDEN • *Department of Medical Pharmacology and Physiology, University of Missouri School of Medicine and Dalton Cardiovascular Research Center, Columbia, MO, USA*

CLIONA O'DRISCOLL • *Hugo Moser Laboratory, Kennedy Krieger Institute, Baltimore, MD, USA; Department of Environmental Health Sciences, Center In Alternatives In Animal Testing, Bloomberg School of Public Health, Johns Hopkins University, Baltimore, MD, USA*

ALEXEI POLIAKOV • *Division of Developmental Neurobiology, MRC National Institute of Medical Research, Mill Hill, London, UK*

JAY D. POTTS • *Biomedical Engineering Program, University of South Carolina, Columbia, SC, USA; Department of Cell Biology and Anatomy, University of South Carolina School of Medicine, Columbia, SC, USA*

ANDREW RAPE • *Department of Biomedical Engineering, Carnegie Mellon University, Pittsburgh, PA, USA*

ANNIE O. SMITH • *Department of Medical Pharmacology and Physiology, University of Missouri School of Medicine and Dalton Cardiovascular Research Center, Columbia, MO, USA*

KATHERINE R. SPEICHINGER • *Department of Medical Pharmacology and Physiology, University of Missouri School of Medicine and Dalton Cardiovascular Research Center, Columbia, MO, USA*

AMBER N. STRATMAN • *Program in Genomics of Differentiation, National Institute of Child Health and Human Development, National Institutes of Health, Bethesda, MD, USA*

ADAM VANDERGRIFF • *Biomedical Engineering Program, University of South Carolina, Columbia, SC, USA*

YU-LI WANG • *Department of Biomedical Engineering, Carnegie Mellon University, Pittsburgh, PA, USA*

SUSAN T. WEINTRAUB • *Department of Biochemistry, San Antonio Cardiovascular Proteomics Center, The University of Texas Health Science Center, San Antonio, TX, USA*

Andriy Yabluchanskiy • *San Antonio Cardiovascular Proteomics Center, The University of Texas Health Science Center, San Antonio, TX, USA; Barshop Institute for Longevity and Aging Studies, The University of Texas Health Science Center, San Antonio, TX, USA; Division of Geriatrics, Gerontology and Palliative Medicine, Department of Medicine, The University of Texas Health Science Center, San Antonio, TX, USA*

Mike T. Yang • *Department of Bioengineering, University of Pennsylvania, Philadelphia, PA, USA*

Jian Zhang • *Department of Biomedical Engineering, Carnegie Mellon University, Pittsburgh, PA, USA*

Chapter 1

Proteomics Analysis of Contact-Initiated Eph Receptor–Ephrin Signaling

Claus Jorgensen and Alexei Poliakov

Abstract

Large-scale biochemical analysis of cell-specific signaling can be interrogated in cocultures of Eph receptor- and ephrin-expressing cells by combining proteomics analysis with cell-specific metabolic labeling. In this chapter, we describe how to perform such large-scale analysis, including the generation of cells stably expressing the receptors and ligands of interest, optimization steps for Eph–ephrin coculture, and the proteomics analysis. As the experimental details may vary depending on the specific system that is being interrogated, the goal of the chapter is mainly to provide sufficient experimental context for experienced researchers to set up and conduct these experiments.

Key words Proteomics, Eph receptors, Ephrins, Cell–cell interactions, Cell signaling

1 Introduction

Reciprocal signaling underlies several aspects of cell communication. This includes direct contact-initiated signaling between neighboring cells, as well as short-range soluble signals. While significant improvements have been made in understanding soluble signals, which can be studied in cellular monocultures, our understanding of contact-dependent signals has been hampered by an inability to study these in coculture. This has primarily been the case with global, unbiased biochemical studies, where information of the cellular origin of the signaling molecule is lost following cellular harvesting, thus preventing cell-specific analysis.

A prototypical example of contact-initiated bidirectional signaling is Eph receptor–ephrin signaling. Eph receptors comprise the largest mammalian receptor tyrosine kinase (RTK) family with 16 receptors and 8 ligands [1–3]. The Eph receptor tyrosine kinases are divided into A- and B-type receptors, based on sequence similarities and affinity for A- and B-type ephrins. The extracellular

Troy A. Baudino (ed.), *Cell-Cell Interactions: Methods and Protocols*, Methods in Molecular Biology, vol. 1066,
DOI 10.1007/978-1-62703-604-7_1, © Springer Science+Business Media New York 2013

region of Eph RTKs contains a highly conserved globular ephrin-binding domain, a cysteine-rich region, two fibronectin type III domains, a sushi domain, and a transmembrane region. The intracellular region contains conserved juxtamembrane tyrosine residues, the tyrosine kinase domain, a sterile alpha motif (SAM) domain, as well as a C-terminal PSD95/Dlg1/ZO-1 (PDZ) binding motif. Importantly, cognate ephrin ligands are also membrane tethered, where A-type ephrins are attached through a glycosylphosphatidylinositol (GPI)-anchor and B-type ephrins are transmembrane. Due to the membrane localization of both the Eph receptors as well as the ephrins, ligand- and receptor-expressing cells need to be juxtaposed to initiate signaling. Following the interaction between an Eph receptor and its cognate ligand, a bidirectional phospho-tyrosine dependent signal is initiated in both cell types [4–6]. More specifically, Eph receptors become phosphorylated on conserved tyrosine residues in the juxta-membrane region as well as in the activation loop of the kinase domain. Subsequently, this leads to the recruitment of SH2 and PTB domain containing proteins such as RASGAP, NCK, and p62DOK [7–9], which results in the remodeling of the cytoskeleton and cell-repulsive behavior. Intriguingly, B-type ephrins are also tyrosine phosphorylated on conserved tyrosine residues in the intracellular segment, which facilitates the association of cognate adaptor-containing proteins.

To study Eph–ephrin signaling, ectodomain fusion proteins have typically been used to simulate their interaction. As such, the extracellular regions of Eph receptors or ephrins have been constructed as fusion protein with the Fc region from IgG, thereby allowing clustered forms of these to be presented to cells of interest. However, since both the Eph receptors and their cognate ephrin ligands initiate a signal, the use of ectodomain fusion proteins for these studies does not necessarily recapitulate the *bidirectional* signal exchanged between the ligand- and receptor-expressing cells. Thus, until recently it was not well understood how the bidirectional signal was regulated between interacting pairs of cells. In this chapter, we describe in detail the experimental setup that was developed to approach this issue.

Briefly, to decipher cell-specific signals we used differential metabolic labeling of receptor- and ligand-expressing cells. As such, cells were labeled with nonradioactive, isotopically enriched, amino acids (stable isotope labeling with amino acids in cell culture, SILAC, [10]), which introduces a known mass shift in all tryptic peptides. Hence, proteomics analysis of mixed populations of labeled receptor- and ligand-expressing cells can be deconvoluted post analysis to provide cell-specific and relative quantification of signaling (*see* Fig. 1 for general workflow). Thus, this workflow enables signals from cocultured cells to be ascribed and quantified in a cell-specific manner.

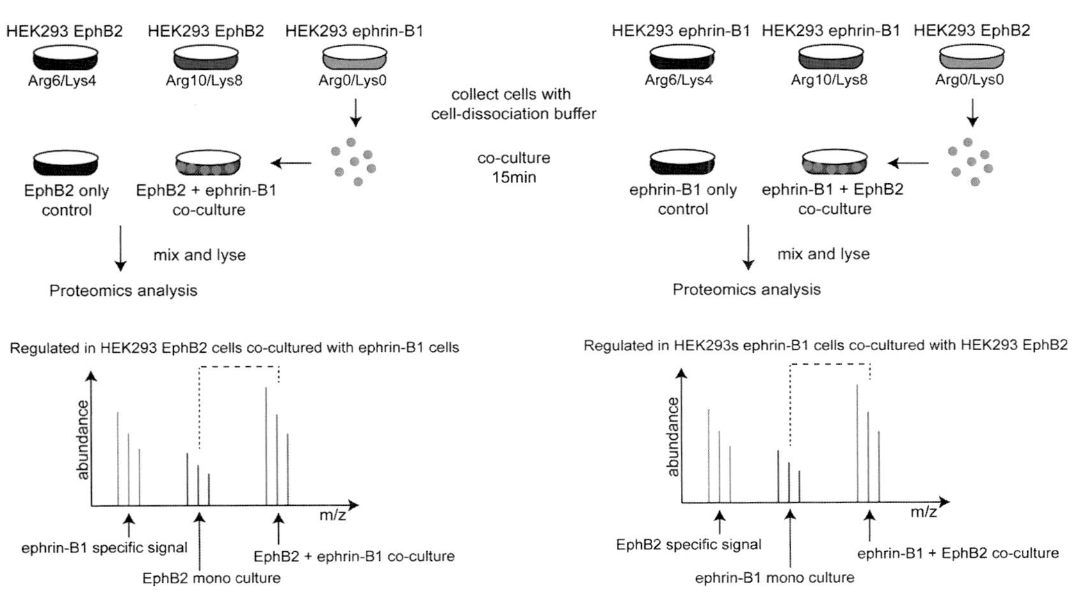

Fig. 1 Outline for the analysis of contact-initiated bidirectional EphB2–ephrin-B1 signaling

2   Materials

2.1 Generation of Stable Eph/Ephrin/Membrane-Targeted GFP-Expressing Cells

1. pCS2+ expression plasmid, mouse EphB2-receptor.

2. pcDNA3 expression plasmid, mouse ephrin-B1.

3. pCS2+ expression plasmid, membrane-targeted GFP [11].

4. Empty pcDNA3 expression plasmid with G418 or hygromycin resistance.

5. G418 and hygromycin.

6. FuGENE 6 (Roche, Bedford, MA, USA).

7. EphrinB1-Fc and EphB2-Fc chimeras (R&D Systems, Minneapolis, MN, USA).

8. Donkey anti-human Cy3-AffiPure F(ab')2 Fragments (Jackson Immunoresearch, West Grove, PA, USA).

9. Donkey anti-human IgG (Jackson Immunoresearch).

10. HEK293 cells (ATCC, Manassas, VA, USA).

11. Culture medium: Dulbecco's modified Eagle's medium (DMEM; high-glucose medium 4.5 g/l without L-glutamine) supplemented with 2 mM L-glutamine, 1 mM sodium pyruvate, and 10 % fetal bovine serum (FBS) Mycoplex.

2.2 Immunoblotting, Immunoprecipitation, and FACS

1. Anti EphB2 antibody [12].

2. Anti ephrin-B1 antibody (C-910, Santa Cruz Biotechnology, Dallas, TX, USA).

3. Anti FLAG antibody (M2 monoclonal, Sigma-Aldrich, St. Louis, MO, USA).

4. Alexa 488 anti-human Fc (Life Technologies, Grand Island, NY, USA).

5. Anti ephrin-B1 phosphotyrosine^{298} antibody (Ab33069, Abcam, Cambridge, MA, USA).

6. Anti-phospho-tyrosine clone 4G10 (EMD Millipore, Billerica, MA, USA).

7. Cell lysis buffer for immunoblotting and immunoprecipitation (PLC buffer): 50 mM Tris–HCl pH 7.5, 1 % Triton X-100, 150 mM NaCl, 1 mM EDTA, 10 % Glycerol, protease inhibitor cocktail, and phosphatase inhibitor cocktail (Sigma).

8. Urea cell lysis buffer for mass spectrometric analysis of protein phosphorylation: 9 M Urea, 20 mM HEPES pH 8.0, and 10 mM NaPPi.

9. Buffers for peptide purification:

   (a) SepPak activation buffer: 100 % Acetonitrile.

   (b) SepPak wash buffer: 0.1 % TFA/deionized water.

   (c) SepPak elution buffer: 60 % Acetonitrile/0.1 % TFA/ deionized water.

10. Phospho-peptide immunoprecipitation (IP) buffer: 100 mM Tris pH 7.4, 100 mM NaCl, and 0.3 % NP40.

11. Phospho-peptide IP wash buffer: 100 mM Tris pH 7.4 and 100 mM NaCl.

12. Peptide elution buffer: 100 mM Glycine, pH 2.5.

13. Anti-phospho-tyrosine^{100} (Cell Signaling Technology, Danvers, MA, USA).

14. Protein A/G Sepharose.

15. Ga(III)-IMAC columns (Pierce, Thermo Scientific, Rockford, IL, USA).

16. Accutase (PAA Laboratories, Dartmouth, MA, USA).

2.3 SILAC Labeling

1. DMEM without Arginine and Lysine pH 7.4 (Caisson Labs, North Loga, UT, USA).

2. Dialyzed FBS.

3. L-Arginine and L-lysine.

4. N15 and C13 enriched L-Arginine (Sigma Isotech: 608033) and L-Lysine (Sigma Isotech: 608041).

5. N15 and C12 enriched L-Arginine (Sigma Isotech: 600113) and L-Lysine (Sigma Isotech: 609021).

6. SILAC labeling medium: DMEM w/o L-Arginine and L-Lysine pH 7.4, 1× Pen/Strep and L-Arginine and L-Lysine added at 50 mg/L.

2.4 Equipment and Software

1. Orbitrap Mass Spectrometer (Orbitrap Classic, Thermo Scientific, Waltham, MA, USA) using an Eksigent nanoLC Ultra 1D plus LC system (AB Sciex, Framingham, MA, USA).

2. Mascot 2.1 (Matrix Science, Boston, MA, USA).

3 Methods

3.1 Generation of Stable Eph/ Ephrin-Expressing HEK293 Cells

For the purpose of this chapter, we describe how stable Eph receptor- and ephrin-expressing cells were generated. HEK293 cells are used to generate stable cell lines expressing (1) EphB2-receptor, (2) EphB2-receptor and membrane-targeted GFP, (3) membrane-targeted GFP as a control, and (4) ephrin-B1 [11]. The advantage of using HEK293 cells is that the endogenous expression of most Eph receptors and ephrin ligands is low. This limits interfering signals exchanged between co-expressed ligands and receptors as well as the possibility of *cis*-interactions. During all the procedures including transfection, clone selection, and expression analysis, cells are cultured at 37 °C with 5 % CO_2 in normal culture medium.

1. Transfection of HEK293 cells are carried out using FuGENE 6 transfection reagent as described by the manufacturer.

2. Perform positive selection of the transfected clones using G418 and/or hygromycin. Initially the concentration of G418 and hygromycin required for selection should be determined experimentally. Specifically, transfect HEK293 cells with an empty vector without antibiotics resistance and leave in culture for 3 days. After 3 days, add G418 or hygromycin to the culture at concentrations ranging from 0.2 to 2.0 mg/ml. The concentration of the antibiotics that kills 100 % of the transfected cells within a week should be chosen for subsequent selection of the clones expressing EphB2, ephrin-B1, and/or membrane-targeted GFP.

3. To generate the mouse EphB2-receptor-expressing cell line, transfect HEK293 cells with the pCS2+ vector containing mouse EphB2 cDNA along with an empty pcDNA3 vector containing the G418 resistance marker (the pcDNA3:pCS2+ ratio should be 1:10).

4. To generate the mouse EphB2-expressing membrane-targeted GFP, transfect clones of EphB2 cells from **step 3** with a pCS2+ vector containing cDNA of GFP fused to the membrane-anchoring signal sequence of GAP-43 [13] along with empty pcDNA3 vector containing hygromycin resistance (the pcDNA3:pCS2+ ratio should be 1:10).

5. To generate the control cell line with membrane-targeted GFP, transfect HEK293 cells with the pCS2+ vector containing cDNA of GFP fused to the membrane-anchoring signal sequence of GAP-43 along with the empty pcDNA vector containing the hygromycin resistance (the pcDNA3:pCS2+ ratio should be 1:10).

6. To generate the mouse ephrin-B1-expressing cell line, transfect HEK293 cells with pcDNA3 containing both mouse ephrin-B1 cDNA and the G418 resistance marker.

7. After selecting 15–20 clones from each transfection group, perform immunoblotting and immunofluorescense to confirm the expression of the EphB2-receptor, ephrin-B1, and membrane-targeted GFP proteins. Also make sure that all proteins are of the correct molecular weight and proper subcellular localization (plasma membrane) (*see* **Note 1**).

3.2 FACS Sorting

1. Plate the EphB2-receptor- and ephrin-B1-expressing cells 24 h before the experiment at 50,000 cells/cm^2.

2. Incubate ephrin-B1-Fc and EphB2-Fc chimeras at a 1:1 molar ratio with Cy3-AffiPure F(ab')2 fragment donkey anti-human IgG for 10 min at 25 °C.

3. Add the formed complexes of fluorescent IgGs bound to EphB2-Fc and ephrin-B1-Fc chimeras at 1 g/ml concentration (based on the Fc-chimeras content) directly to the dishes with ephrin-B1- and EphB2-receptor-expressing cells, respectively.

4. Incubate the cells with the formed protein complexes for 45–60 min to ensure that the fluorescently labeled soluble ligands are completely internalized by the cells [11] (*see* **Note 2**).

5. After labeling of EphB2-receptor- and ephrin-B1-expressing cells, dissociate the cells into a single-cell suspension using Accutase, pellet the cells, and resuspend in ice-cold PBS w/o Mg^{2+}/Ca^{2+} at 1×10^6 cells/ml.

6. Sort the cell suspensions using FACS, collecting the labeled cells that display an intermediate level of intensity (40–60 %).

7. Wash with PBS and plate sorted cells. Culture cells for a week and then use them for subsequent experiments.

3.3 Labeling with Stable Isotopes of Amino Acids in Cell Culture

1. Cells should be grown in normal medium (ArgC12N14/LysC12N14)** medium (ArgC12N15/LysC12N15) or heavy (ArgC13N15/LysC13/N15) labeling medium.

2. Day 1: Seed 200,000 cells onto a 6-well plate.

$**$ Medium labeling medium.

3. Day 2: Remove medium and replace with SILAC labeling medium (light, medium, or heavy).

4. Day 3: Change medium (light to light, medium to medium, and heavy to heavy).

5. Day 4: Split cells 1:5 into labeling medium (light to light, medium to medium, and heavy to heavy).

6. Day 5 and onwards: Change SILAC medium daily and split when approximately 80 % confluent (*see* **Notes 3–5**).

3.4 Coculture Experiment Optimization: Receptor/Ligand Signaling

To initiate contact-dependent bidirectional signaling between Eph receptor- and ephrin-expressing cells, one cell population, for example the Eph receptor-expressing cells, should be plated and allowed to adhere. Subsequently, the other cell population, the ephrin-expressing cells, should be collected and plated onto the adhering Eph receptor-expressing cells. The experiments described here are all aimed to analyze signaling in the *adherent* populations [14].

As there are high levels of expression of the Eph receptor and the ephrin ligand within the stable cell populations, minute amounts of co-expressed ligand or receptor is sufficient to initiate signaling at higher cell densities. Therefore, it is critical to determine the effect of cell density on the level of receptor or ligand activation within the monocultures (*see* **Note 6**). Due to the sensitivity of these experiments, several rounds of optimization should be conducted prior to the proteomics experiments.

1. Day 1: From a dish of approximately 50 % confluent cells, seed Eph- and ephrin-expressing cells at varying levels of confluence (10, 20, 30, and 50 %). It is important to note the number of cells seeded at the individual densities to ensure experimental consistency and that the experiments can be scaled up for the proteomics analysis.

2. Day 2: Serum starve the cells overnight.

3. Day 3: Lyse the cells in PLC buffer and determine the level of activation of the Eph receptor and ephrin ligand by immunoblotting (*see* **Note 7**).

3.5 Coculture Experiment Optimization: Cellular Ratios

Using the settings yielding the lowest level of auto-activation (Subheading 3.4), the next step is to determine the best ratio between Eph receptor- and ephrin-expressing cells for the coculture experiments. These experiments are critical in order to determine the number of ligand-expressing cells required for optimal activation of receptor-expressing cells as well as the number of receptor-expressing cells required for optimal activation of the ligand-expressing cells. *See* Fig. 2 for workflow.

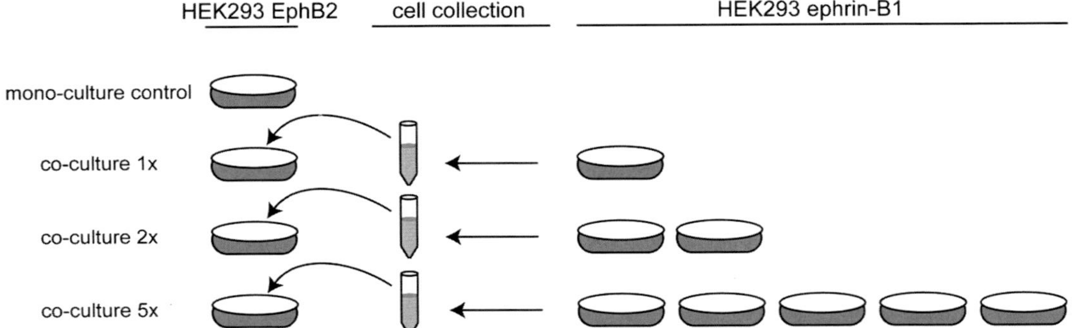

Fig. 2 Workflow for optimization of the cellular ratio between receptor- and ligand-expressing cells. Using an increasing ratio between ephrin-B1- and EphB2-expressing HEK293 cells, the objective is to identify the ratio giving rise to the highest level of receptor activation, thereby ensuring a more homogeneous response

1. Day1: Seed Eph receptor-expressing cells at optimal density for low auto-activation in four 100 mm dishes. In parallel, seed ephrin-expressing cells at their optimal density, but increase the total number of cells seeded to 1×, 2×, and 5× the number of receptor cells seeded onto the 100 mm dish.

2. Day 2: Change the medium and serum starve cells overnight.

3. Day 3: Coculture of cells: Collect ephrin-expressing cells by removing medium and rinsing gently with PBS. Incubate with cell dissociation buffer until cells start to loosen from plate. Use serum-free medium to collect the cells by adding 2–3 ml per plate. Collect cells in 15 ml tubes (1×, 2×, and 5× number of Eph-expressing cells/100 mm dish) and spin at $200 \times g$ for 5 min at room temperature. Remove medium and resuspend cells in 6 ml serum-free medium.

4. Remove medium from Eph receptor-expressing cells. Add 1×, 2×, or 5× ephrin-expressing cells in 6 ml to the Eph receptor-expressing cells and incubate for 15 min at 37 °C.

5. Lyse the cells in PLC buffer and determine the level of receptor activation by immunoblotting (*see* **Note 7**).

In parallel, the optimal ratio of receptor-expressing cells should be determined for optimal activation of ephrin signaling using the same procedure. For these optimization experiments it is critical to determine the number of cells seeded. The reason being that for the subsequent analysis of bidirectional signaling, a larger amount of starting material is required. Thus, in order to ensure consistent results between the optimization experiments and the proteomics experiments cells should be seeded identically.

3.6 Contact-Initiated Signaling for Phospho-Proteomics Analysis

In this section, we describe how we conducted the profiling experiments of phospho-tyrosine-dependent signaling in EphB2- and ephrin-B1-expressing HEK293 cells [14]. Please note that these experiments are built on prior developments as described in studies from Rush et al. and Zhang et al. [21, 22]. It is also worth noting that kits and reagents are now available from Cell Signaling Technology for these experiments.

For the following experiments we describe how ephrin-expressing cells are used to stimulate Eph receptor-expressing cells. Studies in ephrin-expressing cells are conducted in a similar manner, except that Eph receptor-expressing cells are cocultured onto adherent ephrin-expressing cells. All experiments conducted are analytical replicates of four independent biological experiments. It is worth noting that cell contact-initiated signaling in general results in a lower level of activation compared to clustered ectodomain, and thus an increased amount of starting material may be required. For each biological experiment, a total of three 150 mm plates of Eph receptor-expressing cells/label are used, seeded at 2.5×10^6 cells/dish. We have found that adding double the amount of ligand-expressing cells to the receptor-activating cells works well in our hands [14].

1. Day 0: Seed Eph receptor- and ephrin-expressing cells at the predetermined density.

2. Eph-expressing cells are medium and heavy labeled, whereas ephrin-expressing cells are light labeled.

3. Day 1: Serum starve both Eph receptor- and ephrin-expressing cells.

4. Day 2: Aspirate medium from the ephrin-expressing cells and rinse gently with PBS. Aspirate the PBS and add 2–3 ml of cell dissociation buffer to each plate. Incubate until cells begin to loosen from the plate. Collect the cells in serum-free medium and spin down at $200 \times g$ for 5 min. Resuspend the cells in the appropriate volume of medium (8 ml/150 mm dish of cell; *see* **Note 8**), count the cells, and add the appropriate number of cells as previously determined. Leave the cells in the incubator for the amount of time of interest.

5. After incubation, remove medium and lyse the cocultured cells (heavy labeled Eph-expressing cells and light labeled ephrin-expressing cells) in urea lysis buffer.

6. Remove medium from the medium labeled Eph receptor-expressing cells and lyse an equal amount of Eph receptor-expressing cells in the same urea lysis buffer.

7. All cells should be lysed in a total volume of 10 ml of urea lysis buffer. All cell harvesting should be carried out at room temperature, as the urea will precipitate out on ice.

8. Following cell lysis, vortex the samples twice for 30 s and shear DNA by sonication. The sample should be sonicated three times at 15 W for 15 s.

9. Clear samples by centrifugation at $10,000 \times g$ for 10 min at room temperature. Transfer the cleared lysate to a new tube and determine protein concentration of the sample using the BCA assay according to the manufacturer's instruction. This will determine the amount of trypsin added later and will aid in troubleshooting if the signal is low.

3.7 Reduction, Alkylation, and Digestion

1. Reduce the sample by adding 5 mM DTT final concentration and incubate at room temperature for 45 min. It is important not to incubate at higher temperatures, as this will increase protein carbamylation.

2. Add 100 mM iodoacetamide (final concentration) and incubate in the dark at room temperature for 20 min.

3. Dilute the sample to 2 M urea and 20 mM HEPES pH 8.0, final concentration.

4. Take out a predigestion sample (20–50 µl) and add proteomics-grade trypsin at a final amount of 1:50 enzyme/substrate ratio. Leave the sample to digest for 24 h with rotation at room temperature.

5. Check the digestion efficiency by SDS-PAGE/Coomassie staining, comparing pre- and post-digested samples to ensure that all proteins have been digested completely. If non-digested bands appear on the gel, check pH is 8, then add extra trypsin and leave the sample to digest for an additional 24 h. Stop the digestion by acidification with TFA to a final concentration of 1 %.

3.8 Peptide Purification

1. Prepare a SepPak C18 column by attaching it to a 20 ml syringe and place over a beaker (to collect the flow-through).

2. Activate the C18 reverse-phase column by adding 5 ml of 100 % acetonitrile. In parallel, clear the acidified samples by centrifugation at $2,000 \times g$ for 10 min at room temperature. Transfer the cleared lysate to a new tube. This step is critical to remove any debris and precipitated peptides; failure to clear the sample sufficiently will lead to clogging of the SepPak column.

3. Following activation of the SepPak columns by 100 % acetonitrile, equilibrate the column by addition of 7 ml of 0.1 % TFA. After flow through, repeat with the addition of 7 ml of 0.1 % TFA. Let the buffer empty by gravity flow.

4. Load the sample into the syringe and let it empty by gravity flow. Air bubbles may form at the junction between the SepPak cartridge and the syringe, but they can be gently removed by flicking to facilitate flow.

5. After sample loading, rinse the syringe with deionized water, reassemble the syringe and SepPak column, and wash the sample twice with 12 ml of 0.1% TFA.

6. Elute the sample with 8 ml 60 % acetonitrile/0.1 % TFA and collect in a new polypropylene tube.

7. Divide the eluted samples into two 4 ml aliquots, snap freeze the eluted samples, and dry down in a speed vac. Lyophilized peptides can be stored at −20 °C.

3.9 Phospho-Peptide Enrichment

1. Resuspend one vial of lyophilized peptides (corresponding to 10–20 mg of total protein depending on coculture conditions) in peptide IP buffer. Incubate on ice for 30 min and vortex frequently to solubilize the peptides.

2. Determine the pH of the solution and adjust to pH 7.4 (on ice) with 2 M Tris pH 9.2 and vortex to bring the peptides into solution.

3. Once the solution has reached a pH of 7.4, centrifuge the solution at 15,000 rpm at 4 °C for 20 min and transfer the supernatant to a clean tube. Collect a small aliquot for normalization (Subheading 3.13).

4. Add 20 μl pTyr100 antibody and incubate at 4 °C with rotation for 6 h.

5. Add 20 μl packed Protein A/G IP08 beads and incubate for an additional 10 h.

6. Collect the beads by centrifugation at 3,000 rpm for 30 s. Transfer the supernatant to a clean tube and add 3 μg 4G10 and incubate at 4 °C with rotation for 6 h.

7. Add 20 μl packed Protein A/G IP08 beads and incubate for an additional 10 h.

8. Wash beads three times in 1 ml IP wash buffer and elute peptides using 100 μl 1 M glycine pH 2.5 at room temperature with rotation.

9. Collect the eluted peptides and further enrich for phospho-peptides using Ga(III)-IMAC.

3.10 IMAC Enrichment of Phospho-Peptides

Due to nonspecific interactions between highly abundant non-phosphorylated peptides and the agarose resin used for immunoprecipitation, it is recommended to further enrich the phosphorylated peptides. For these studies we used phospho-peptide isolation kit from Pierce according to the manufacturer's instruction (*see* **Note 9**), but alternative phospho-peptide isolation methods may also be employed [22, 23].

1. Eluted samples are added to the Ga(III)-IMAC resin, mixed by flicking the tube gently, and incubated at room temperature for 20 min.

2. Collect samples by centrifugation at $1,000 \times g$ for 1 min. Remove the flow through and save in a fresh tube.

3. Wash the resin with 50 µl 0.1 % acetic acid.

4. Wash the resin twice with 50 µl 0.1 % acetic acid/10 % acetonitrile.

5. Wash the resin in 75 µl water (LC-MS grade).

6. Elute the sample by incubating the IMAC resin with 50 µl 100 mM ammonium bicarbonate pH 9.5 for 15 min at room temperature, followed by centrifugation at $1,000 \times g$ for 1 min and collection of the flow through.

7. Repeat **step 6** twice.

8. Adjust samples to pH 2.5/3.0 with 5 % formic acid. It is critical to adjust the pH to acidic conditions rapidly after elution to prevent loss of phospho-peptides under alkaline conditions.

3.11 Mass Spectrometry Analysis and Data Analysis

In this section, we describe how phospho-peptide samples are analyzed by mass spectrometry, but mainly focus on describing our experience with the data analysis and normalization of bidirectional signaling. For the experiments described above, heavy labeled Eph receptor-expressing HEK293 cells were mixed with light labeled ephrin-expressing HEK293 cells, whereas medium labeled Eph-expressing HEK293 cells were left untreated and act as a control. Therefore, comparing the relative abundances (using the extracted ion currents) between heavy and medium labeled peptides reflects the regulation of the signaling between cocultured and control Eph receptor-expressing cells.

1. Analyze isolated phospho-peptides on an Orbitrap Mass Spectrometer using an Eksigent nanoLC Ultra 1D plus LC system. Samples are analyzed at a flow rate of 100 nl/min, with the following gradient 1 % MeCN/0.1 % FA, 120 min: 25 % MeCN/0.1 % FA, 140 min: 40 % MeCN/0.1 % FA, 160 min: 80 % MeCN/0.1 % FA, 170 min: 1 % MeCN/0.1 % FA.

2. Operate the mass spectrometer in a data-dependent acquisition mode with one scan at 60,000 resolution followed by selection of the three most intense peptides for MS/MS in the linear ion trap (IT-CID).

3. Search the data using with the following parameters: precursor mass accuracy, 12 ppm, MS/MS accuracy, 0.6 Da, with the following modifications: cysteine alkylation (fixed), phospho-STY (variable), SILAC labeled amino acids (arginine and lysine) (variable), methionine oxidation (variable), and NQ deamidation (variable).

4. Perform all quantification using MSquant; however, other quantification tools are available such as Maxquant (http://maxquant.org/) [24, 25], Proteome Discoverer (http://portal.thermobrims.com/), and Mascot (http://www.matrixscience.com/home.html).

3.12 Determination of Labeling Efficiency

A critical step in developing an optimal platform for the analysis of bidirectional signaling is determining the efficiency whereby the isotopically labeled amino acids are incorporated. To evaluate the incorporation of medium and heavy labeled amino acids, samples from cells grown in labeling medium (passages 3–6) are analyzed by LC-MS.

1. Seperate 30 μg of total cell lysate by SDS-PAGE as per lab protocols.

2. Stain the gel with Colloidal Brilliant Blue and excise one or two representative bands for in-gel digestion [26].

3. Identify the peptides by LC-MS.

Peptides from these analyses should contain the labeled amino acids, and the ratio between the labeled and unlabelled forms can be used to determine labeling efficiency. We typically obtain efficiencies above 95 %. Importantly, these analyses should also be used to evaluate the extent of arginine-to-proline conversion. This is a critical control, as arginine-to-proline conversion will impact the accuracy of peptide quantification.

3.13 Normalization

For each experiment examining bidirectional signaling, it is strongly recommended to evaluate the level of heavy to medium labeled peptides prior to phospho-peptide enrichment. This is critical, as the ratio between these labeled peptides reflects the accuracy of cell mixing, and it is imperative that this is controlled for. In particular, since the relative level between heavy and medium labeled phospho-peptides is used to quantify the extent of signaling, inconsistencies in cell mixing will impact the accuracy of quantification.

To evaluate cell mixing, analyze the sample of digested peptides that were aspirated prior to adding the anti-phospho-tyrosine antibody (Subheading 3.11). Assuming near-complete labeling efficiency and that equal amounts of heavy and medium labeled cells were plated for the coculture experiments, these ratios should be very close to 1:1. If the ratio is unexpectedly off, it is suggested that labeling efficiency and cell mixing be reevaluated.

4 Notes

1. In our experience, minute differences in the level of expression of Eph receptors and ephrin ligands is a source of experimental variation. Also, we have noticed that the cell clones gradually lose the expression of EphB2 and ephrin-B1 over time, even when grown in the presence of the selection antibiotics. This is a critical point, as the cells are subsequently used in functional cell sorting assays and biochemical analysis

of bidirectional signaling [11]. Consistent readout in these assays heavily depends on the fraction of cells expressing EphB2-receptor or ephrin-B1 before mixing them together. Therefore, to eliminate cells with low expression levels of Eph receptor and ephrin-ligand and to ensure that level of expression of the overexpressed proteins is homogeneous across the cell populations, we sort the cell lines by FACS. This ensures that the sorted cell populations are 100 % positive for EphB2-receptor and ephrin-B1 and can then be used in subsequent experiments for up to 8–10 passages without loss of expression.

2. As a result, the cells expressing either EphB2-receptor or ephrin-B1 will become labeled with Cy3 fluorescent probe and non-expressing cells will remain nonfluorescent. The amount of internalized complexes of soluble ligands should be proportional to the amount of the EphB2-receptors or ephrin-B1 on the cell surface. Therefore, the fluorescent intensity of the cells should reflect the level of overexpressed proteins in the cell population.

3. For labeling of other cell lines and for a more detailed description of SILAC labeling, the reader is referred to www.silac.org and the following articles [15, 16].

4. Frequent changes of the SILAC labeling medium can be used to increase the speed of cell labeling [27].

5. We recommend determining the labeling efficiency from cells collected after 3–6 passages in SILAC labeling medium to determine when the labeling efficiency is >95 % and whether arginine-to-proline conversion has taken place [17–19]. We have not observed any significant arginine-to-proline conversion in the HEK293s that we have used; however should this be the case, inclusion of proline (C12N14) in the labeling medium can suppress this effect [20].

6. Expression of high levels of Eph receptors does lead to increased auto-activation and therefore a certain level of receptor phosphorylation is to be expected.

7. The level of ephrin-B1 auto-activation can be directly determined by immunoblotting against phosphorylated intracellular tyrosine residues. For EphB2, no commercial antibodies exist for analysis of the phosphorylated receptor, and therefore the receptors should be immunoprecipitated prior to blotting with a phospho-tyrosine antibody. However, due to the level of conservation between receptors, phosphospecific antibodies targeting other members of the Eph receptor family may cross-react. This should be validated experimentally.

8. The volume of seeding medium during the coculture experiments can have effects on experimental variation. We would recommend seeding in a reduced volume, sufficient to cover the dish. This will ensure a higher degree of simultaneous interaction between receptor- and ligand-expressing cells.

9. The phospho-peptide purification kit from Pierce (#89853) was recently discontinued. There are however alternatives available from Pierce (#88300), Sigma (P-9740), and GL science (www.glsciencesinc.com).

Acknowledgements

The authors would like to acknowledge the contribution of Tony Pawson and David Wilkinson. Claus Jorgensen is a recipient of a CR-UK Career Establishment Award.

References

1. Pasquale EB (2008) Eph-ephrin bidirectional signaling in physiology and disease. Cell 133:38–52

2. Lackmann M, Boyd AW (2008) Eph, a protein family coming of age: more confusion, insight, or complexity? Sci Signal 1:re2

3. Pasquale EB (2010) Eph receptors and ephrins in cancer: bidirectional signaling and beyond. Nat Rev Cancer 10:165–180

4. Holland SJ, Gale NW, Mbamalu G et al (1996) Bidirectional signalling through the EPH-family receptor Nuk and its transmembrane ligands. Nature 383:722–725

5. Bruckner K, Pasquale EB, Klein R (1997) Tyrosine phosphorylation of transmembrane ligands for Eph receptors. Science 275:1640–1643

6. Henkemeyer M, Orioli D, Henderson JT et al (1996) Nuk controls pathfinding of commissural axons in the mammalian central nervous system. Cell 86:35–46

7. Holland SJ, Gale NW, Gish GD et al (1997) Juxtamembrane tyrosine residues couple the Eph family receptor EphB2/Nuk to specific SH2 domain proteins in neuronal cells. EMBO J 16:3877–3888

8. Wybenga-Groot LE, Baskin B, Ong SH et al (2001) Structural basis for autoinhibition of the Ephb2 receptor tyrosine kinase by the unphosphorylated juxtamembrane region. Cell 106:745–757

9. Cowan CA, Henkemeyer M (2001) The SH2/SH3 adaptor Grb4 transduces Bephrin reverse signals. Nature 413:174–179

10. Ong S-E, Blagoev B, Kratchmarova I et al (2002) Stable isotope labeling by amino acids in cell culture, SILAC, as a simple and accurate approach to expression proteomics. Mol Cell Proteomics 1:376–386

11. Poliakov A, Cotrina ML, Pasini A et al (2008) Regulation of EphB2 activation and cell repulsion by feedback control of the MAPK pathway. J Cell Biol 183:933–947

12. Henkemeyer M, Marengere LE, McGlade J et al (1994) Immunolocalization of the Nuk receptor tyrosine kinase suggests roles in segmental patterning of the brain and axonogenesis. Oncogene 9:1001–1014

13. Moriyoshi K, Richards LJ, Akazawa C et al (1996) Labeling neural cells using adenoviral gene transfer of membrane-targeted GFP. Neuron 16:255–260

14. Jørgensen C, Sherman A, Chen GI et al (2009) Cell-specific information processing in segregating populations of Eph receptor ephrin-expressing cells. Science 326:1502–1509

15. Ong S-E, Mann MA (2007) Practical recipe for stable isotope labeling by amino acids in cell culture (SILAC). Nat Protocol 1:2650–2660

16. Harsha HC, Molina H, Pandey A (2008) Quantitative proteomics using stable isotope labeling with amino acids in cell culture. Nat Protocol 3:505–516

17. Blagoev B, Mann M (2006) Quantitative proteomics to study mitogen-activated protein kinases. Methods 40:243–250

18. Van Hoof D, Pinkse MW, Oostwaard DW et al (2007) An experimental correction for arginine-to-proline conversion artifacts in SILAC-based quantitative proteomics. Nat Methods 4:677–678

19. Park SK, Liao L, Kim JY et al (2009) A computational approach to correct arginine-to-proline conversion in quantitative proteomics. Nat Methods 6:184–185

20. Bendall SC, Hughes C, Stewert MH et al (2008) Prevention of amino acid conversion in SILAC experiments with embryonic stem cells. Mol Cell Proteomics 7:1587–1597

21. Rush J, Moritz A, Lee KA et al (2005) Immunoaffinity profiling of tyrosine phosphorylation in cancer cells. Nat Biotechnol 23:94–101

22. Zhang Y, Wolf-Yadlin A, Ross PL et al (2005) Time-resolved mass spectrometry of tyrosine phosphorylation sites in the epidermal growth factor receptor signaling network reveals dynamic modules. Mol Cell Proteomics 4:1240–1250

23. Thingholm TE, Jensen ON, Robinson PJ et al (2007) SIMAC (sequential elution from IMAC), a phosphoproteomics strategy for the rapid separation of monophosphorylated from multiply phosphorylated peptides. Mol Cell Proteomics 7:661–671

24. Cox J, Matic I, Hilger M et al (2009) A practical guide to the MaxQuant computational platform for SILAC-based quantitative proteomics. Nat Protocol 4:698–705

25. Cox J, Mann M (2008) MaxQuant enables high peptide identification rates, individualized p.p.b.-range mass accuracies and proteome-wide protein quantification. Nat Biotechnol 26:1367–1372

26. Shevchenko A, Tomas H, Havlisbreve J et al (2007) In-gel digestion for mass spectrometric characterization of proteins and proteomes. Nat Protocol 1:2856–2860

27. Boisvert FM, Ahmad Y, Gierlinski M et al (2012) A quantitative spatial proteomics analysis of proteome turnover in human cells. Mol Cell Proteomics 11:M111.011429

Chapter 2

Control of Vascular Tube Morphogenesis and Maturation in 3D Extracellular Matrices by Endothelial Cells and Pericytes

George E. Davis, Dae Joong Kim, Chun-Xia Meng, Pieter R. Norden, Katherine R. Speichinger, Matthew T. Davis, Annie O. Smith, Stephanie L.K. Bowers, and Amber N. Stratman

Abstract

An important advance using in vitro EC tube morphogenesis and maturation models has been the development of systems using serum-free defined media. Using this approach, the growth factors and cytokines which are actually necessary for these events can be determined. The first model developed by our laboratory was such a system where we showed that phorbol ester was needed in order to promote survival and tube morphogenesis in 3D collagen matrices. Recently, we have developed a new system in which the hematopoietic stem cell cytokines, stem cell factor (SCF), interleukin-3 (IL-3), and stromal derived factor-1α (SDF-1α) were added in conjunction with FGF-2 to promote human EC tube morphogenesis in 3D collagen matrices under serum-free defined conditions. This new model using SCF, IL-3, SDF-1α, and FGF-2 also works well following the addition of pericytes where EC tube formation occurs, pericytes are recruited to the tubes, and vascular basement membrane matrix assembly occurs following EC–pericyte interactions. In this chapter, we describe several in vitro assay models that we routinely utilize to investigate the molecular requirements that are critical to EC tube formation and maturation events in 3D extracellular matrix environments.

Key words Endothelial cells, Pericytes, 3D matrices, Cell–cell interaction, Vascular tube morphogenesis

1 Introduction

Many recent studies have addressed the molecular basis for how blood vessels form and mature [1–7]. This work has progressed through the development of novel experimental approaches and the utilization of both in vivo and in vitro models. Animal models used to investigate tissue vascularization include those in mice, zebrafish, and avian species and also, sophisticated in vitro models have been developed to mimic vessel assembly including lumen and tube formation as well as sprouting behavior and maturation

Troy A. Baudino (ed.), *Cell-Cell Interactions: Methods and Protocols*, Methods in Molecular Biology, vol. 1066,
DOI 10.1007/978-1-62703-604-7_2, © Springer Science+Business Media New York 2013

17

events such as endothelial cell–mural cell interactions leading to vascular basement membrane matrix formation (Fig. 1). It is being increasingly appreciated that the parallel use of both approaches is leading to more rapid discoveries to elucidate the underlying mechanisms that control these fundamental cellular events. For example, blood vessel assembly involves important homotypic interactions between endothelial cells (ECs) as well as heterotypic EC interactions with mural cells such as pericytes and vascular smooth muscle cells and critical signal transduction cascades that result from such cell–cell contacts [1, 5, 7].

Major advances in recent years have occurred in our understanding of the molecular mechanisms underlying EC tube formation in 3D matrices and much of these advances have occurred due to the use of in vitro 3D morphogenesis assay systems in collagen or fibrin matrices [1, 6–9] (Fig. 1). Vascular tube formation occurs secondary to integrin signaling, cell surface proteolysis, and signaling through Rho GTPases and protein kinase cascades [1, 2, 6, 10]. Interestingly, a key step in EC tubulogenesis is the generation of vascular guidance tunnels [11] (Figs. 2 and 3) which are matrix-free spaces that are created as a result of MT1-MMP-mediated proteolysis. This proteolytic process occurs in coordination with integrin, Rho GTPase, as well as protein kinase C, Src family kinase, Pak kinase, Raf, and Erk kinase-dependent signaling [12–15]. Following creation of vascular guidance tunnels, which are matrix templates in 3D matrices, EC motility and tube remodeling events occur within these spaces and importantly, mural cells such as pericytes are recruited to EC-lined tubes on their abluminal surface [5, 11, 16, 17] (Figs. 2 and 3). Thus, after pericyte recruitment to tubes, both ECs and pericytes reside within tunnel spaces and both cell types work together to assemble the vascular basement membrane [5, 16] (Figs. 2 and 3). Disruption of pericyte recruitment to EC tubes in vitro or in vivo leads to markedly reduced basement membrane matrix deposition, showing that EC–mural cell interactions play a major role in stimulating this key extracellular matrix remodeling process [5, 17].

An important advance using in vitro EC tube morphogenesis and maturation models has been the development of systems using serum-free defined media [2, 5, 7, 10, 18, 19]. Using this approach, the growth factors and cytokines which are actually necessary for these events can be determined. The first model developed by our laboratory was such a system where we showed that phorbol ester was needed in order to promote survival and tube morphogenesis in 3D collagen matrices [18]. We added both FGF-2 and VEGF in this model, but clearly phorbol ester addition was the major factor that allowed the system to function. More recently, we have developed a new system in which the hematopoietic stem cell cytokines, stem cell factor (SCF), interleukin-3 (IL-3), and stromal derived factor-1α (SDF-1α) were added in

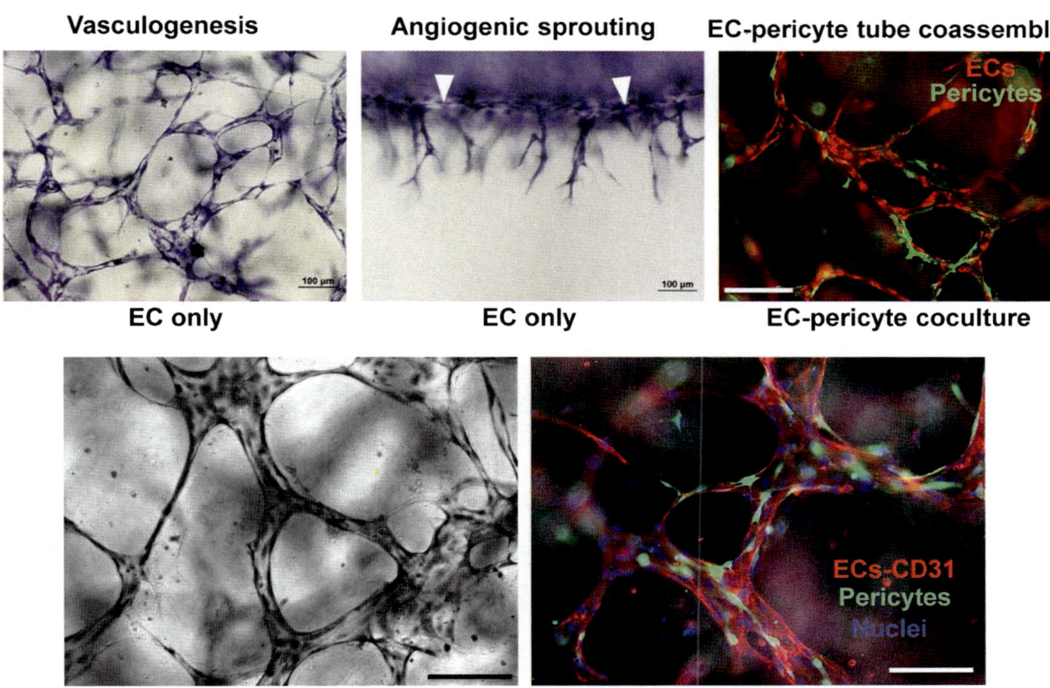

Fig. 1 Microassay systems of human EC tube morphogenesis, sprouting, and EC–pericyte tube co-assembly using serum-free defined media in 3D collagen matrices. In all cases, the hematopoietic stem cell cytokines, SCF, IL-3, and SDF-1α were used in conjunction with FGF-2 to obtain these marked morphogenic responses and these factors were polymerized into the collagen matrix at the start of the assay. An assay is illustrated showing vasculogenic tube assembly (*upper left*) (ECs only) whereby single ECs are seeded together and over time assemble into tube networks in 3D collagen matrices. Cultures were fixed at 72 h, stained with toluidine blue, and photographed. An angiogenic sprouting assay is illustrated (*upper middle*) whereby the hematopoietic cytokines as well as FGF-2 and VEGF-165 were added into the matrix and ECs were seeded onto the gel surface. After 24 h, the culture was fixed, stained with toluidine (*blue*), cross-sectioned, and photographed. *Arrow-heads* indicate the monolayer surface. EC–pericyte tube co-assembly assays are illustrated in the *upper right panel* and *lower panels*. In the *upper right panel* experiment, mCherry-labeled ECs were seeded with GFP-labeled pericytes and after 72 h, the culture was fixed and photographed under fluorescence. In the *lower panel experiment*, unlabeled ECs were primed with VEGF-165 and FGF-2 and then were trypsinized and mixed with GFP-labeled pericytes in 3D collagen matrices which contained SCF, IL-3, SDF-1α, and FGF-2. After 96 h, the cultures were fixed and stained with toluidine blue (*lower left panel*) or were immunostained with antibodies to CD31 (*lower right panel*) and then photographed. The latter culture was also stained with Hoechst dye to label nuclei. Bar = 100 µm

conjunction with FGF-2 to promote human EC tube morphogenesis in 3D collagen matrices under serum-free defined conditions [16, 20] (Fig. 1) (and in the absence of phorbol ester). Interestingly, the addition of VEGF and FGF-2 in combination fails to promote tube formation under these defined conditions [20]. This new model using SCF, IL-3, SDF-1α, and FGF-2 also works well following the addition of pericytes where EC tube

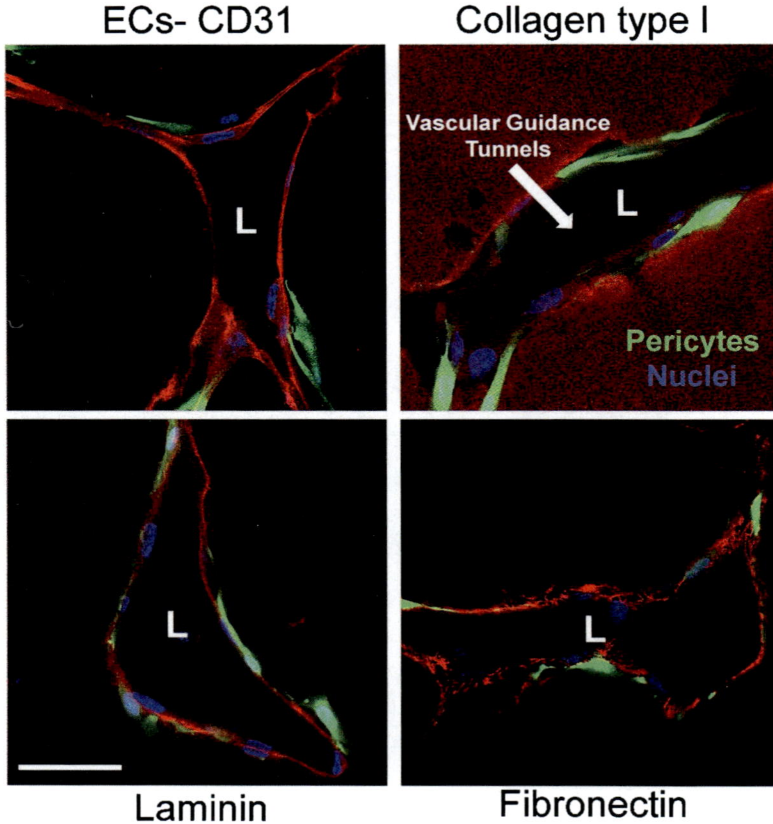

Fig. 2 EC–pericyte tube co-assembly occurs within vascular guidance tunnels and leads to vascular basement membrane matrix assembly between the ECs and pericytes. Unlabeled ECs and GFP-pericytes were cocultured in 3D collagen matrices for 96 h and were immunostained for the indicated molecules, CD31, collagen type I, laminin, and fibronectin (red staining). These cultures were also stained with Hoechst dye to stain nuclei. For the ECM protein stains, no detergent was added to ensure that only extracellular staining was being observed. Representative confocal sections are shown for each stain. *Arrow* indicates a vascular guidance tunnel space. *L* indicates lumen space. Bar = 25 μm

formation occurs, pericytes are recruited to the tubes, and vascular basement membrane matrix assembly occurs following EC–pericyte interactions [16, 20] (Figs. 1 and 2). We also recently demonstrated that VEGF and FGF-2 (or their combination) can prime ECs for subsequent EC tube morphogenic responses to the hematopoietic cytokines [20] (Fig. 1, lower panels). This priming effect occurs in part secondary to upregulation of hematopoietic cytokine receptors on ECs including the SCF receptor, c-Kit, the IL-3 receptor (IL-3 receptor α), and the SDF-1α receptor, CXCR4 [20]. Thus, in this defined model of human EC tube morphogenesis and sprouting, VEGF's primary action appears to be that of a

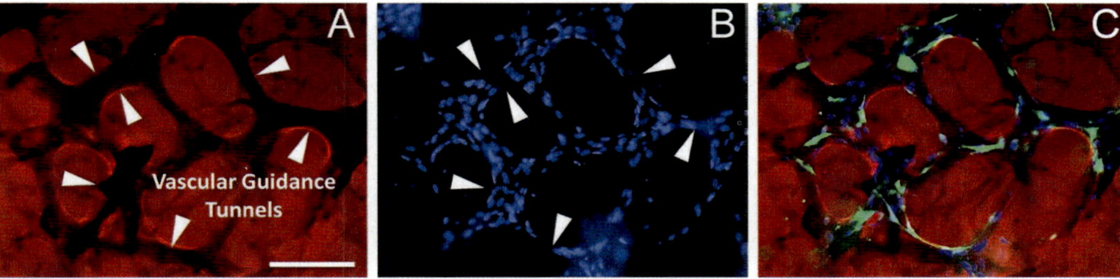

Fig. 3 Vascular tube maturation events resulting from EC–pericyte interactions occur within vascular guidance tunnels in 3D extracellular matrices. ECs create vascular guidance tunnels secondary to MT1-MMP-mediated proteolysis in coordination with the EC lumen and tube formation process. Pericytes are recruited to EC-lined tubes and within vascular guidance tunnels as illustrated in this figure. Unlabeled ECs were seeded with GFP-pericytes in 3D collagen matrices and after 96 h of EC–pericyte tube co-assembly, cultures were fixed and stained with anti-collagen type I antibodies (*red*) to delineate vascular guidance tunnels and Hoechst dye to label nuclei (*blue*). Note that all of the cells were mixed randomly in the gel at the start of the assay, but after 96 h, all of the cells are present within vascular guidance tunnels as demonstrated by the nuclear stain showing that the ECs are assembled together in tube structures and that pericytes have all recruited to the abluminal surface of these tubes. *Arrowheads* indicate the borders of vascular guidance tunnels. Bar = 100 µm

primer which acts to prepare ECs for morphogenic responses to subsequent growth factor/cytokine and extracellular matrix signals [20]. In this chapter, we describe several in vitro assay models that we routinely utilize to investigate the molecular requirements that are critical to EC tube formation and maturation events in 3D extracellular matrix environments.

2 Materials

The key materials necessary to perform these morphogenic assays include human endothelial cells and human pericytes which are grown as described [8, 21]. Also, we can label these cells with fluorescent labels including green fluorescent protein (GFP) or monomeric Cherry (mCherry) that are introduced using recombinant lentiviruses. We perform morphogenic assays utilizing rat tail collagen type I 3D matrices and use recombinant growth factors and cytokines which are added either into the collagen matrix, the culture media, or both. All of the assays are performed in half-area 96-well plates. Immunostaining is performed by fixing the 3D collagen gels and staining the gels using methods similar to whole-mount staining of tissues or embryos. Cultures can also be stained with dyes such as toluidine blue which facilitates our ability to assess and quantitate lumen and tube formation by photography and tracing tube areas using Metamorph software [8].

2.1 Endothelial Cell and Pericyte Culture

1. Human umbilical vein endothelial cells (HUVECs) are obtained from Lonza (Basel, Switzerland) and are grown from passages 2–6.

2. HUVEC medium: Medium 199 containing 20 % fetal calf serum (FCS), 400 μg/ml of bovine hypothalamic extract, and 100 μg/ml of heparin (Supermedia; *see* **Notes 1–3**).

3. Tissue culture flasks.

4. Phosphate-buffered solution (PBS).

5. 0.1 % Gelatin in PBS.

6. Human brain vascular pericytes (HBVP) are obtained from ScienCell (Carlsbad, CA, USA) and are grown from passages 2–10.

7. Bovine retinal pericytes are cultured from bovine retinas as previously described [15].

8. Pericyte medium: Dulbecco's modified Eagle's medium (DMEM) with 10 % FCS.

9. Collagen type I is purified from rat tails as described [8] and is lyophilized from a 0.1 % acetic acid solution in deionized water (sterile filtered). The collagen is then resuspended in 0.1 % acetic acid in water solution at a final concentration of 7.1 mg/ml. We obtain approximately 100 mg of collagen type I from one rat tail (*see* **Note 1**).

10. Reduced serum supplement II (RSII): This supplement is made by mixing a combination of insulin, transferrin, selenium, and fatty acid-free bovine serum albumin with added C18 oleic acid which is prepared as described [8]. The mixture is frozen and stored at −20 °C.

11. Recombinant growth factors, cytokines, and other medium additives: Recombinant FGF-2 (EMD Millipore, Billerica, MA, USA), SCF, IL-3, and SDF-1α (R&D Systems, Minneapolis, MN, USA) and ascorbic acid (Sigma-Aldrich, St. Louis, MO, USA).

12. Half-area 96-well tissue culture plates (A/2) (Costar, Corning, Tewksbury, MA, USA).

2.2 Lentivirus Production

1. Lentiviruses are generated using the ViraPower Lentiviral Expression system (Invitrogen, Grand Island, NY, USA).

2. We utilize a lentiviral construct (pLenti6/V5 TOPO) with a blasticidin resistance gene or we use the Lenti-X system (Clontech, Mountain View, CA, USA) that uses either the lentiviral vector pLVX-IRES-Neo (G418 selection) or pLVX-IRES-Puro (puromycin selection).

3. A lentiviral vector packaging system (Clontech) is utilized along with 293FT cells (grown in DMEM with 10 % FCS) to make recombinant lentiviruses as described by the manufacturer.

3 Methods

3.1 EC Tube Morphogenesis Assay in 3D Collagen Matrices

1. Trypsinize confluent human ECs, wash 1× with 10 ml of M199, and then resuspend at 1×10^7 cells/ml (gently mix cells with a P200 tip to break up small clumps).

2. Seed cells at 2×10^6 cells/ml in 2.5 mg/ml of collagen type I suspended in M199. To make 1 ml of collagen gel, 350 µl of 7.1 mg/ml type I collagen in 0.1 % acetic acid, 39 µl of 10× M199, 2.1 µl of 5 N NaOH, and 409 µl of 1× M199 are mixed together thoroughly.

3. Add 200 µl of ECs to the collagen gel mixture (**step 2**), which is then swirled and placed on ice.

4. The recombinant growth factors, SCF, IL-3, SDF-1α, and FGF-2, are all added at 200 ng/ml within the collagen matrices.

5. Add 28 µl of gel per well and periodically tap the plates on each edge to make certain that the gels are evenly distributed in each well prior to polymerization.

6. Allow the plates to equilibrate in a CO_2 incubator for 30 min and then add 100 µl of culture medium to each well. The culture medium is M199 which contains a 1:250 dilution of RSII supplement, 40 ng/ml of FGF-2, and 50 µg/ml of ascorbic acid.

7. We add 125 µl of water in every well surrounding the gelatin-containing wells and we also add 150 µl of water in non-well areas that surround each of the wells to maintain humidity and to reduce potential dehydration of culture wells over time.

8. Allow assays to proceed for 1, 3, or 5 days and at these time points, cultures can be fixed with either 2 % paraformaldehyde in PBS or 3 % glutaraldehyde in PBS. Paraformaldehyde-fixed gels can be utilized to perform immunostaining using various antibodies such as CD31 (Figs. 1 and 2) or extracellular matrix proteins (Fig. 2). Glutaraldehyde-fixed gels are typically stained with 0.1 % toluidine blue in water, which is an excellent stain to visualize tubes (Fig. 1).

3.2 EC Tube Sprouting Assay in 3D Collagen Matrices

1. Collagen gels (2.5 mg/ml) are prepared as above (Subheading 3.1) and contain 200 ng/ml of recombinant IL-3, SDF-1α, SCF, FGF-2, and VEGF-165.

2. Add 28 µl of gel to each well in 96-well plates.

3. Allow gels to polymerize and equilibrate the pH by placing them in a CO_2 incubator for 30–60 min.

4. Seed ECs at 50,000 cells/well in M199 culture medium (100 µl/well) that contains a 1:250 dilution of RSII, as well as FGF-2 at 40 ng/ml and ascorbic acid at 50 µg/ml.

5. Allow cultures to incubate for 1, 3, or 5 days of culture and after this time, fix the cultures with 3 % glutaraldehyde in PBS (140 μl per well). For the 5-day culture, 60 μl of medium is removed and replaced with fresh medium at 3 days of culture and it is prepared as described above. Fixed cultures are stained with 0.1 % toluidine blue in water. Gels can be bisected with a clean razor blade to visualize a cross section of EC sprouting and tube morphogenesis (Fig. 1).

3.3 EC–Pericyte Tube Co-assembly Assay in 3D Collagen Matrices

1. Plate ECs in collagen gels (2.5 mg/ml) at 2×10^6 cells/ml, and GFP-labeled pericytes at 0.4×10^6 cells/ml.

2. The gels also contain 200 ng/ml of SCF, IL-3, SDF-1α, and FGF-2 and add 28 μl of the cell–gel mixture to the A/2 microwells.

3. After polymerization and equilibration in a CO_2 incubator for 30–60 min, add M199 culture medium (100 μl/well), which also contains a 1:250 dilution of RSII, 40 ng/ml of FGF-2, and 50 μg/ml of ascorbic acid.

4. Allow cultures to proceed for 3 or 5 days of culture. For the 5-day culture time point, cultures are fed on day 3 by removing and replacing 60 μl of medium with fresh medium.

5. Cultures are fixed with either 2 % paraformaldehyde in PBS for immunostaining and fluorescence microscopy or 3 % glutaraldehyde in PBS to then stain with 0.1 % toluidine blue in water.

3.4 Priming of ECs with VEGF Isoforms to Activate EC Morphogenic Responses

1. Culture ECs in T25 or T75 cm² flasks to confluence in Supermedia as described above (Subheading 2.1) and then wash 2× with 5 or 15 ml of media, respectively.

2. Culture the cells for 16–20 h in M199 medium containing a 1:250 dilution of RSII, VEGF-165 at 40 ng/ml, and FGF-2 at 40 ng/ml.

3. ECs are then trypsinized and used for either the EC-only or EC–pericyte coculture assays as described above (Subheading 3.3). In all cases, EC morphogenic responses are strongly enhanced using this VEGF priming protocol. Our work suggests that a major morphogenic influence of VEGF is to prime or prepare ECs for morphogenic responses that are stimulated by the hematopoietic stem cell cytokines, SCF, IL-3, and SDF-1α [20].

3.5 Analysis of Vascular Basement Membrane Deposition Resulting from EC–Pericyte Interactions During Tube Co-assembly in 3D Collagen Matrices

We have shown that a key consequence of pericyte recruitment to EC-lined tubes is the deposition of vascular basement membrane matrices [5, 16]. We have demonstrated that both cell types directly contribute to the deposition process and have further shown that EC-only cultures fail to deposit a vascular basement membrane matrix under our defined media conditions in 3D collagen matrices [16]. For this analysis, we have utilized either EC–pericyte cocultures with human or bovine pericytes and both cell types work very well for this analysis.

EC–pericyte cocultures are established in 3D collagen gels as described above (Subheading 3.3) and after 3 or 5 days of culture, cultures are fixed with 2 % paraformaldehyde (for immunofluorescence staining) or 3 % electron microscopy-grade glutaraldehyde in culture media (for transmission electron microscopy). To fully demonstrate that basement membrane matrix assembly has occurred, both immunofluorescence microscopy and transmission electron microscopy need to be performed. Another critical point is that the immunostaining protocol for vascular basement membrane matrix assembly is performed in the absence of detergent such that only extracellular staining is observed [16]. We typically utilize unlabeled ECs and GFP-pericytes and thus stain for the vascular basement membrane components, laminin, fibronectin, collagen type IV, nidogen 1, nidogen 2, and perlecan, using AlexaFluor 594-conjugated secondary antibodies (Molecular Probes) so that they are red in color (Fig. 2). To stain ECs, we typically utilize antibodies to CD31 and utilize the AlexaFluor-conjugated secondary antibodies as illustrated in Fig. 2. Nuclei are stained with Hoechst dye (Figs. 2 and 3).

3.6 Labeling ECs or Pericytes with Membrane-Targeted GFP or mCherry Facilitates Visualization of EC Tube Morphogenesis and EC–Pericyte Tube Co-assembly in 3D Collagen Matrices

An important advance in our ability to image vessel tube morphogenesis and maturation events is to label ECs or pericytes with fluorescent markers. A requirement for such labels is that they should have no deleterious influence on their functional ability to participate in morphogenic events in 3D matrices. This technology has been particularly useful in real-time imaging of EC–pericyte tube co-assembly, where we demonstrated for the first time how the two cells interact with each other during these events in 3D extracellular matrices [5, 16, 17]. We have successfully labeled pericytes with enhanced GFP and these cells appear to be functionally normal in our assay models allowing us to readily quantitate pericyte motility and recruitment to EC-lined tubes in 3D matrices. More recently, we have labeled them with mCherry constructs and they also appear to be functionally normal. For ECs, we have found that membrane-targeted GFP or mCherry constructs work best in the assays described above. For this purpose, we have utilized a Ras membrane-targeting sequence that is fused to the C-terminus of AcGFP (Ac-GFP-F) (Clontech) and made a recombinant lentivirus. We have also created an mCherry-F lentiviral construct that has the same membrane targeting motif fused to the C-terminus of mCherry. The sequences of the primers used for this construction are:

AcGFP-F Upstream-Not1

5′-AGGCGGCCGCACCATGGTGAGCAAGGGCGCCG
AGCTGTTCAC-3′

AcGFP-F Downstream-BamH1

5′-AGGGATCCTCAGGAGAGCACACACTTGCAGCTC-3′

mCherry-F Upstream-Not1

5′-AGGCGGCCGCACCATGGTGAGCAAGGGCGAG
GAGG-3′

mCherry-F Downstream-BamH1, 5′-AGGGATCCTCAGGAGA
GCACACACTTGCAGCTCATGCAGCCGGGG
CCACTCTCATCAGGAGGGTTCAGCT
TCTTGTACAGCTCGTCCATGCC-3′

We have cloned these constructs into a pLVX-IRES-Neo lentiviral vector (Clontech), infected ECs or pericytes, and then selected for the cells carrying the vector using G418 at 200 μg/ml. Assays with fluorescent ECs or pericytes are established in the same manner as with unlabeled cells.

3.7 Real-Time Imaging of EC Tubulogenesis and EC–Pericyte Tube Co-assembly in 3D Collagen Matrices

1. Establish EC-only or EC–pericyte cocultures in 3D collagen matrices in half-area 96-well microwell plates as described above (Subheading 3.3).

2. Place a glass plate with the dimensions of 75 mm × 50 mm on the surface of the 96-well plates and then replace the lid to the plate. We have found that the glass plate strongly decreases potential condensation on the top of the plates which can interfere with imaging.

3. We utilize two independent microscopy systems for our real-time imaging, which are the Nikon Eclipse TE2000-E with Photometrics CoolSNAP-HQ2 camera and the Leica DMI6000B with Hamamatsu ORCA-ER camera which are each equipped with an incubator that surrounds the microscopic stage to control the temperature of the cultures to 37 °C. In addition, there is a CO_2 controller that is placed on the surface of the 96-well plates that delivers CO_2 to a level of 5 % on the cultures.

4. Image cultures at different starting times and collect images every 10 min for 24, 48, or 72 h. This can be adjusted to individual experiments. We can acquire light, fluorescent, or both images depending on the particular purpose of the experiment and movie. We usually acquire images every 10 min and after this time, files are converted into movies using Metamorph

software. In many cases, we overlay these images with others obtained at a different fluorescent wavelength (i.e., ECs carry mCherry while pericytes carry GFP) (Fig. 1). After this step and construction of movies using Metamorph, they are converted into Windows Media Player or Quicktime files for routine viewing and publication purposes.

4 Notes

1. Two key reasons for our success over the years in performing these EC tube morphogenesis assays is that we make our own bovine brain extract as well as rat tail collagen type I preparations. In this way, we have an internal consistency (for more than 15 years) that is not dependent on variable commercial sources or availability issues.

2. For our assays to be optimal, the HUVEC cells should be of passages 2–6 and the growth media needs to be utilized exactly as we describe. Because of different additives that are used in commercially available media, they may adversely impact the performance of the assays that we describe. This issue has not been assessed in detail; however, our assays are highly reproducible and if performed as we describe, they should work well each time an assay is established.

3. Our assay systems are highly compatible with siRNA suppression protocols for either ECs or pericytes which we have described previously in detail [8].

Acknowledgements

This work was supported by NIH grants, HL79460, HL87308, and HL105606, to GED. We also wish to thank the efforts of our previous excellent technicians who helped with this work including Anne Mayo, Rachel Mahan, and Kristine Malotte.

References

1. Davis GE, Stratman AN, Sacharidou A et al (2011) Molecular basis for endothelial lumen formation and tubulogenesis during vasculogenesis and angiogenic sprouting. Int Rev Cell Mol Biol 288:101–165

2. Davis GE, Koh W, Stratman AN (2007) Mechanisms controlling human endothelial lumen formation and tube assembly in three-dimensional extracellular matrices. Birth Defects Res C Embryo Today 81:270–285

3. Xu K, Cleaver O (2011) Tubulogenesis during blood vessel formation. Semin Cell Dev Biol 22:993–1004

4. Sacharidou A, Stratman AN, Davis GE (2012) Molecular mechanisms controlling vascular lumen formation in three-dimensional extracellular matrices. Cells Tissues Organs 195:122–143

5. Stratman AN, Davis GE (2012) Endothelial cell-pericyte interactions stimulate basement

membrane matrix assembly: influence on vascular tube remodeling, maturation, and stabilization. Microsc Microanal 18:68–80

6. Iruela-Arispe ML, Davis GE (2009) Cellular and molecular mechanisms of vascular lumen formation. Dev Cell 16:222–231

7. Davis GE (2012) Molecular regulation of vasculogenesis and angiogenesis: recent advances and future directions. In: Homeister JW, Willis MS (eds) Molecular and translational medicine. Springer, New York, NY, pp 169–206

8. Koh W, Stratman AN, Sacharidou A et al (2008) In vitro three dimensional collagen matrix models of endothelial lumen formation during vasculogenesis and angiogenesis. Meth Enzymol 443:83–101

9. Nakatsu MN, Hughes CC (2008) An optimized three-dimensional in vitro model for the analysis of angiogenesis. Meth Enzymol 443: 65–82

10. Davis GE, Bayless KJ, Mavila A (2002) Molecular basis of endothelial cell morphogenesis in three-dimensional extracellular matrices. Anat Rec 268:252–275

11. Stratman AN, Saunders WB, Sacharidou A et al (2009) Endothelial cell lumen and vascular guidance tunnel formation requires MT1-MMP-dependent proteolysis in 3-dimensional collagen matrices. Blood 114:237–247

12. Koh W, Sachidanandam K, Stratman AN et al (2009) Formation of endothelial lumens requires a coordinated PKCepsilon-, Src-, Pak- and Raf-kinase-dependent signaling cascade downstream of Cdc42 activation. J Cell Sci 122:1812–1822

13. Sacharidou A, Koh W, Stratman AN et al (2010) Endothelial lumen signaling complexes control 3D matrix-specific tubulogenesis through interdependent Cdc42- and MT1-MMP-mediated events. Blood 115:5259–5269

14. Bayless KJ, Davis GE (2002) The Cdc42 and Rac1 GTPases are required for capillary lumen formation in three-dimensional extracellular matrices. J Cell Sci 115:1123–1136

15. Saunders WB, Bohnsack BL, Faske JB et al (2006) Coregulation of vascular tube stabilization by endothelial cell TIMP-2 and pericyte TIMP-3. J Cell Biol 175:179–191

16. Stratman AN, Malotte KM, Mahan RD et al (2009) Pericyte recruitment during vasculogenic tube assembly stimulates endothelial basement membrane matrix formation. Blood 114:5091–5101

17. Stratman AN, Schwindt AE, Malotte KM et al (2010) Endothelial-derived PDGF-BB and HB-EGF coordinately regulate pericyte recruitment during vasculogenic tube assembly and stabilization. Blood 116:4720–4730

18. Davis GE, Camarillo CW (1996) An alpha 2 beta 1 integrin-dependent pinocytic mechanism involving intracellular vacuole formation and coalescence regulates capillary lumen and tube formation in three-dimensional collagen matrix. Exp Cell Res 224:39–51

19. Davis GE, Stratman AN, Sacharidou A (2011) Molecular control of vascular tube morphogenesis and stabilization: regulation by extracellular matrix, matrix metalloproteinases and endothelial cell-pericyte interactions. In: Gerecht S (ed) Biophysical regulation of vascular differentiation. Springer, New York, pp 17–47

20. Stratman AN, Davis MJ, Davis GE (2011) VEGF and FGF prime vascular tube morphogenesis and sprouting directed by hematopoietic stem cell cytokines. Blood 117:3709–3719

21. Davis GE, Camarillo CW (1995) Regulation of endothelial cell morphogenesis by integrins, mechanical forces, and matrix guidance pathways. Exp Cell Res 216:113–123

Chapter 3

Analyzing Cell–Cell Interactions in 3-Dimensional Adhesion Assays

Stephanie L.K. Bowers and Troy A. Baudino

Abstract

The organization of cells is key to the proper formation and function of tissues and it appears to be dependent upon various intracellular and extracellular signals. These signals come from cell–cell interactions, as well as interactions with the surrounding extracellular milieu. In order to investigate these properties and interactions among cells, our lab utilizes and has developed several techniques that provide a 3-dimensional, in vivo-like environment for in vitro cell culture. In this chapter, we describe several techniques for isolating primary cardiac cells, including myocytes, endothelial cells, and fibroblasts. In addition, we discuss and outline an adhesion assay and an aggregation assay that can be used for numerous cell types, as well as a collagen gel assay for examination of cell–cell and cell–matrix interactions.

Key words Cell–cell interactions, Cardiac fibroblasts, Endothelial cells, Myocytes, Tube formation, Cardiac cell isolation

1 Introduction

Investigating cell behavior is a key element in acquiring information about any biological system or disease. Gene and protein expression of a particular cell population is paramount to its behavior and the function of the cells in a given tissue or organ. In vitro experiments are utilized to focus on a specific signaling pathway and help to break down exactly how a cell population contributes to its microenvironment and, ultimately, its biological relevance. These types of studies have provided a basis for much of what we know about gene and protein expressions of individual cell populations. However, it would also be beneficial to learn more about how cell behavior may change in response to the influence of other cell types, especially during disease processes where individual cell populations may fluctuate (e.g., during cancer growth, scar formation after myocardial infarction, and organ development). At present, many in vitro studies only focus on one particular cell population, without direct investigation of how interactions and communication with

Troy A. Baudino (ed.), *Cell-Cell Interactions: Methods and Protocols*, Methods in Molecular Biology, vol. 1066, DOI 10.1007/978-1-62703-604-7_3, © Springer Science+Business Media New York 2013

surrounding cells may affect gene expression, protein expression, and behavior of the cell population of interest. Isolating signaling pathways that result from heterogeneous cell–cell interactions provides a more natural, in vivo-like environment for the study of cell behavior and function, and holds a wealth of information for the development of more targeted treatments and therapies.

One consideration, and complication, for studying cell–cell interactions in vitro is the origin of the cell population being studied. Growing evidence has shown that a given cell population does not necessarily confer the same phenotype and behavior for every tissue in which it resides [1, 2]. To this end, it would be prudent to use cells derived from the same tissue of the disease being studied or the same tissue in which the study model is generated, as these cells will provide the most relevant information possible.

In vitro experiments involving multiple cell populations are mostly hampered by cell isolation techniques, and providing a 3-dimensional (3D) environment in which to study cell–cell interactions. The expansion of cell-specific markers for developing antibodies and new forms of isolation techniques have helped tremendously. Sterile fluorescence-activated cell sorting (FACS) is optimal, but involves expensive equipment and technical experience that is not always readily available or easy to obtain. Here, we describe methods involving magnetic bead-conjugated antibodies to isolate endothelial cells (ECs) and fibroblasts from mouse hearts. In addition, we describe several 3D assays that can be used to study interactions between cardiac myocytes, fibroblasts, and ECs: an adhesion assay, aggregation assay, and collagen gels. The cell adhesion and aggregation assays were originally developed in our lab to examine cardiac myocyte and fibroblast interactions [3]; however, more recently we have used these assays to examine cardiac fibroblast and endothelial cell interactions. The collagen gel assay has been modified from a system using human-derived cells (developed in the lab of George Davis [4]), to be amenable to study lumen and tube formation using rodent cells. This collagen gel assay is preferred over commercially available substrates, such as Matrigel, since other basement membrane proteins and growth factors are absent in this assay, making it possible to extract more information about the signals involved in vascularization and angiogenesis in tissues.

2 Materials

2.1 Endothelial Cell and Fibroblast Isolation from Mouse Hearts

Everything used should be kept sterile. Prepared Dynabead–antibody conjugates should be used within 1–2 weeks of preparation for best results. Antibodies used can be replaced with those from other companies, but the yield and purity of the isolation depend on the quality of the antibody.

1. Laminin (20 μg/mL) and 0.1 % gelatin-coated T-25 and T-75 tissue-culture treated flasks for plating cells. Gelatin may be used for both fibroblasts and ECs, but ECs tend to plate and grow better on laminin-coated plates (*see* **Note 1**).

2. Cell isolation medium: Dulbecco's modified Eagle's medium (DMEM; 4.5 g/L glucose) + 20 % fetal bovine serum (FBS) + 100 IU penicillin and 100 μg/ml streptomycin (1× Pen/Strep). Set aside an aliquot to keep cold for initial heart isolation, but warm the rest of the isolation medium for culture.

3. EC growth medium: DMEM (4.5 g/L glucose) + 20 % FBS + 100 μg/ml Heparin + 100 μg/ml Endothelial Cell Growth Supplement (ECGS; BD Biosciences, San Jose, CA, USA) + 1× Pen/Strep + 1× nonessential amino acids (NEAA) + 1× L-GLUTAMINE. Pre-warm for cell culture.

4. Fibroblast growth medium: DMEM (4.5 g/L glucose) + 10 % FBS + 1× Pen/Strep + 1× NEAA + 1× L-glutamine.

5. Tissue digestion medium: Type I Collagenase (Worthington Biochemical Corp., Lakewood, NJ, USA; 2–2.5 mg/ml (~200 U/mg)) in isolation medium. Sterile filter using 0.2 μm filter and pre-warm at 37 °C before use (*see* **Note 2**).

6. Dulbecco's phosphate buffer solution (DPBS) without Ca^{2+} and Mg^{2+} for incubation with conjugated antibodies (*see* **Note 3**).

7. Instruments for animal dissection (autoclaved/sterilized): Two scissors and two forceps (one of each used for the initial incisions, the other for handling the heart in the thoracic cavity and excision).

8. Materials for tissue dissociation (sterile): Forceps, 2 ml square-bottom microcentrifuge tube, long-tipped scissors (that can comfortably reach into the microcentrifuge tube), 16-gauge metal cannulae, 30 ml syringe, 3 ml syringe, and cell strainer (70 μm pore size).

9. Other materials needed: 10 ml Petri dish for cleaning the heart, 5 ml polystyrene snap-cap tubes (used during antibody incubation), magnetic separator (Invitrogen, various types/sizes), and 50 ml conical tubes (*see* **Note 4**).

10. Antibodies: Purified Rat anti-Mouse CD31 (PECAM-1, clone MEC13.3) antibody (BD Biosciences, #553369) for primary EC sorting; Purified Rat anti-Mouse CD102 (ICAM-2) antibody (BD Biosciences, #553325) for secondary EC sorting; Purified anti-rabbit DDR2 (Santa Cruz Biotechnology, Dallas, TX, USA) for fibroblasts.

11. Dynabeads (*see* **Note 5**): Sheep anti-Rat IgG (Invitrogen, Grand Lisland, NY, USA, #11035) for CD31 and CD102 (ECs); M-280 Sheep anti-Rabbit IgG (Invitrogen, #11203D) for DDR2 (fibroblasts) (*see* **Note 6**).

2.2 Myocyte and Fibroblast Isolation from Neonatal Rat Hearts

All reagents and solutions should remain sterilized or be sterile filtered before use. Pre-warming solutions is also necessary to improve cell recovery and integrity.

1. 80 Neonatal rat pups, 1–3 days old.

2. 1× HBSS (dilute from 10× HBSS with autoclaved deionized water).

3. Enzyme solution for 80 neonatal rat pups: 300 ml of 1× HBSS, 3.2 mg Trypsin (Worthington Biochemical Corp., 2–2.5 mg/ml (~200 U/mg)), 23.2 mg alpha-Chymotrypsin (Sigma-Aldrich, St. Louis, MO, USA), and 40 mg Collagenase II (Invitrogen).

4. Dispersion medium: Dissolve 1 pouch of 1× Medium 199 (M199) powder in 800 ml of deionized water, add 2.2 g sodium bicarbonate, pH to 7.35–7.4, and fill up to 1,000 ml with deionized water. Dissolve 1 pouch of DMEM low-glucose powder in 800 ml of deionized water, add 7.4 g sodium bicarbonate, pH to 7.35–7.4, and fill up to 1,000 ml with deionized water. Measure 500 ml of M199 and add it to the 1,000 ml of DMEM. Sterile filter the remaining 500 ml of M199 and store at 4 °C for future use. Measure 425 ml of the M199/DMEM mixture and sterile filter it into a 500 ml bottle. Add 50 ml of horse serum, 25 ml of FBS, and 1× Pen/Strep. Prepare another 500 ml of dispersion media if needed, sterile filter any remaining M199/DMEM mixture, and store at 4 °C.

5. Percoll gradient supplies: Percoll (GE Healthcare Life Sciences, Piscataway, NJ, USA), sterile deionized water, 12.5× Ads solution (116 mM NaCl, 20 mM HEPES, 1 mM NaH_2PO_4, 5.5 mM glucose, 5.4 mM KCl, and 0.8 mM $MgSO_4$). Add to 90 ml deionized water, pH to 7.35, and fill to 100 ml total. Sterile filter and store at 4 °C.

   (a) For 80 pups, prepare "Bottom" and "Top" mixtures for gradient centrifugation as follows: Bottom—15.88 ml of Percoll, 2.08 ml of 12.5× Ads, 8.04 ml of sterile deionized water; Top—10.64 ml of Percoll, 2.08 ml of 12.5× Ads, 13.28 ml of sterile deionized water.

6. Other supplies needed: Water-jacketed stir flask (125–250 ml volume) connected to a circulating water bath maintained at 37 °C; 150 mm tissue culture dish and biohazard bag for pup sacrifice; two sets of sterile scissors and forceps (one set for pup sacrifice and one for cleaning and chopping hearts); razor blade for chopping hearts; two 100 mm dishes with 10–20 ml of 1× HBSS for pumping and cleaning hearts; two 100 mm dishes with 10 ml of 1× HBSS for chopping hearts, and an extra 5 ml of 1× HBSS for rinsing dishes (25 ml of 1× HBSS total to start digestion).

2.3 3-Dimensional Collagen Gel Assay

All steps and reagents should remain sterile. All reagents used to make the collagen gel should remain on ice until polymerization is desired; the gel will congeal quickly at room temperature.

1. Use trypsin:EDTA to collect isolated cells (passages 1–3) for experiment.

2. Collagen gel mixture: For 1 ml of 1.7 mg/ml collagen (low concentration), combine 350 µl of 5 mg/ml Type I rat-tail collagen (Invitrogen), 39 µl of 10× M199, 2.1 µl of 5 N NaOH, and 409 µl of 1× M199. For 1 ml of 2.68 mg/ml collagen (high concentration), combine 525 µl of 5 mg/ml type I collagen, 58.5 µl of 10× M199, 3.15 µl of 5 N NaOH, and 213 µl of 1× M199 (*see* **Note 7**).

3. 50 ml conical tube and/or 1.5 ml microcentrifuge tube, depending on the amount of collagen gel needed and the experimental groups.

4. 96 Half-area well clear flat bottom tissue culture-treated microplate (Corning Life Sciences, Tewksbury, MA, USA) (*see* **Note 8**).

5. Feeding medium: DMEM (4.5 g/L glucose) + 10 % FBS + 1× NEAA + 1× L-GLUTAMINE + 1× Pen/Strep (*see* **Note 9**).

6. Growth factors: Use FGF-2 (100 ng/ml) with or without VEGF-165 (100 ng/ml) for a positive control (Upstate Biotechnology, EMD Millipore, Billerica, MA, USA), or to induce lumen and tube formation.

3 Methods

Carry out all procedures under a sterile laminar flow hood, except for animal dissection and heart excision.

3.1 Endothelial Cell and Fibroblast Isolation from Mouse Hearts

3.1.1 Preparation of Dynabeads (Prepare at least 1 Day Prior to Isolation)

1. Resuspend a working amount of Dynabeads in a microcentrifuge tube containing 1 ml of Buffer 1 (PBS + 0.1 % BSA + 2 mM EDTA), depending on the number of isolations planned. Mix well by pipetting, but take care not to lose volume/beads in the pipette tip (*see* **Note 10**).

2. Place the tube on a magnetic separator and leave for 2 min. Remove the supernatant by pipette or decanting—do not aspirate. Repeat this wash with Buffer 1 three times.

3. Resuspend beads to their original volume with Buffer 1 (i.e., if 500 µl of beads were initially resuspended in **step 1**, use 500 µl of Buffer 1 here).

4. Add 5 µl of purified Ab for each 100 µl of Dynabeads used for CD31 (primary sort) or CD102 (secondary sort) (stock antibody concentration is 0.5 mg/ml). Adjust the volume according to stock antibody concentration if necessary.

Incubate overnight on a rotator at 4 °C (*see* **Note 11**). In the morning, repeat washes in **steps 1–3** (4×) with Buffer 1.

5. Resuspend beads in original volume with Buffer 1 to maintain beads at their original concentration. Store beads at 4 °C and use within 1–2 weeks (*see* **Note 12**).

3.1.2 Dissection and Heart Excision

1. Sacrifice the mouse using an IACUC-approved method (e.g., CO_2 or isoflurane inhalation overdose).

2. Spray and wipe down the mouse with 70 % ethanol; pin/tape to a dissecting tray if desired. Using clean scissors and forceps, carefully pull up on the xiphoid process, make a small incision through the peritoneum and abdominal muscles, and then cut down the sides of the body to expose the diaphragm, and then up through the chest cavity; take care to gently cut the diaphragm away from the pericardium.

3. Using different sterile scissors and forceps, cut the heart away from the thoracic wall and thymus, and remove the heart from the chest cavity by pulling up and cutting at the aorta (*see* **Note 13**). Place into a Petri dish with 15–20 ml of ice-cold cell isolation medium. Pump the heart gently to remove the blood, and cut it in half so that it is completely submerged in the medium (*see* **Note 14**).

4. Repeat **steps 1** and **2** for a second mouse, and place in the same Petri dish. Move contents to a sterile hood.

3.1.3 Tissue Dissociation

1. Place the heart pieces from both mice in a 2 ml microcentrifuge tube containing approximately 500 µl of pre-warmed tissue digestion medium.

2. Mince the pieces with scissors for <1 min, being careful not to scrape the tube sides too much. Pour and pipette the pieces into 25 ml of pre-warmed tissue digestion medium, rinsing the tube with media to ensure collection of all cells (*see* **Note 15**).

3. Incubate at 37 °C with gentle agitation for 45–60 min, or until most pieces are fully digested. For agitation, use an orbital shaker (80 rpm) placed in an incubator, a shaking water bath, or inversion in a hybrid oven. Alternatively, briefly invert/mix the suspension by hand every 5 min during incubation.

4. Once during the incubation, use a 30 ml syringe attached firmly to a 16-gauge cannula or blunt-end needle to triturate the suspension 5–7 times, taking care to avoid frothing of the cells (*see* **Note 16**).

5. At the end of the incubation, triturate again 5–7 times: if chunks are still visible, incubate for an additional 10–15 min, but no longer than 75 min total (*see* **Note 17**).

6. Allow any remaining small chunks to settle. Pipette the cell suspension through a 70 μm disposable cell strainer into a new 50 ml conical tube. Wash the tube and strainer with 10 ml of cell isolation medium.

7. Spin down the cell suspension at $400 \times g$ (1,300 rpm in a GH3.7 rotor) for 8 min at 4 °C. Resuspend the pellet in 1 ml of DPBS (*see* **Note 3**). Leave behind any stubborn clumps (*see* **Note 18**).

8. Count nucleated cells to use as a reference point for subsequent isolations (*see* **Note 19**). Transfer the 1 ml cell suspension to a sterile 5 ml round-bottom snap-cap polystyrene tube.

3.1.4 Sterile Cell Sorting

1. Add 15 μl of CD31-conjugated Dynabeads, and push the cap of the tube on all the way. Incubate on a rotator (20 rpm) at room temperature for 15 min (*see* **Note 20**).

2. Mount the tube in a magnetic separator under the sterile laminar flow hood and leave for 2 min. For maximum recovery, let the majority of beads bind, and then quickly tilt the tube upside down to get any beads that may be stuck in the cap (cap should still be on tight from rotation). Flick and gently loosen the tube cap, getting as much of the suspension down into the tube. Remove the cap, and add a small amount of medium to rinse the sides of the tube.

3. Remove the supernatant via pipette. The beads and cells should look somewhat foamy if there is good yield of positive selection. Collect all wash supernatant (flow through) to spin down after all subsequent rinses.

4. Remove tube from the magnetic separator and resuspend beads and cells in 2.5 ml of cell isolation medium by vigorous trituration with a 16-gauge cannula attached to a 3 ml syringe. Unless trituration is vigorous, contaminating cells and clumping will be evident in all cultures. Mount in magnetic separator for 2 min.

5. Repeat **steps 2–4** until the supernatant is clear (4–5 washes).

6. Resuspend ECs and beads in EC growth medium and plate onto one gelatin-coated T-25 cell culture flask.

7. The next day, gently rinse the flask twice with cell isolation medium to remove any loosely adherent cells, and replace with fresh EC growth media (*see* **Note 21**).

8. If sorting for fibroblasts or another cell type, centrifuge the collected wash flow through, and continue by repeating **steps 3–7**, plating fibroblasts in fibroblast growth medium on a 35 mm dish to start with and assess the yield (*see* **Notes 6** and **22**).

3.1.5 Second Sort for Endothelial Cells

1. Feed cells with EC growth medium every other day. Let cells grow until approaching confluence at 3–5 days after initial plating. There should be monolayer groups of ECs comprising the majority of the cells. Other mesenchymal cells are notable for the way they pile up and do not form monolayers.

2. Detach cells with trypsin:EDTA by rinsing with DPBS, incubating with trypsin:EDTA for approximately 1 min or only as long as it is needed to detach cells. Add 10 ml of cell isolation medium to inactivate trypsin and rinse flask.

3. Transfer cell suspension to a 15 ml tube and spin down at $400 \times g$ for 8 min. Remove supernatant and resuspend cell pellet in 1 ml of DPBS.

4. Add 15 µl of ICAM-2/CD102-coated Dynabeads to the cell suspension. Incubate for 15 min at room temperature. Place on a magnetic sorter and leave for 2 min. Remove supernatant. Wash gently twice in 2.5 ml of media with mild trituration.

5. After the final wash, resuspend cells in 10 ml of EC growth medium and plate on a gelatin-coated T-75 tissue culture flask (*see* **Note 23**).

3.1.6 Cell Propagation

1. Feed with EC or fibroblast growth medium every other day.

2. Split cells 1:3 with trypsin:EDTA when they reach 80 % confluency (*see* **Note 24**).

3.2 Myocyte and Fibroblast Isolation from Neonatal Rat Hearts

All procedures should be performed under a sterile hood. Dispersion media may be prepared prior to isolation; the enzyme solution, however, should be prepared within 1 h of beginning the isolation.

1. Sacrifice rat pups by scruffing at the nape of the neck to stabilize the body and quickly decapitate. Cut down the midline sternum to open the chest and place the heart in the first 100 mm rinse dish with 10–20 ml of 1× HBSS. Pump blood out of heart and remove atria. After approximately 40 pups, switch to fresh 1× HBSS to rinse.

2. Transfer heart to a second 100 mm collection dish and chop ventricles with scissors; leave in larger chunks while sacrificing the remaining pups, and chop hearts into small pieces with a sterile razor blade after all hearts are collected.

3. Add chopped hearts and all of 1× HBSS to the water-jacketed stir flask, and rinse dishes with 5 ml of 1× HBSS for a total of 25 ml. Add 25 ml of warm enzyme solution.

4. Stir at 400–500 rpm (higher rpm is better, but do not introduce many bubbles or frothing) for 12 min. Turn stir plate off and allow tissue to settle; discard supernatant (decant or pipette).

5. Add 60 ml of fresh enzyme solution and stir for 15 min.

6. Allow tissue to settle, and collect enzyme solution (decant or pipette) in two 50 ml conical tubes with 5 ml of dispersion medium each. Return any tissue to the stir flask that may have been transferred. Add 60 ml of fresh enzyme solution to the tissue and repeat the 15-min incubation.

7. During enzyme incubation, spin collected 50 ml tubes at $250 \times g$ at room temperature for 10 min, aspirate and discard supernatant, and resuspend pellet in 5 ml dispersion medium; place in incubator with loose cap to allow oxygen diffusion.

8. Repeat **steps 5–7** until the heart pieces are very light colored and/or enzyme solution is used up (not all tissue will be digested). After the first spin of collected cells, combine subsequent resuspended pellets in a single 50 ml conical, and keep in incubator.

9. During spins, prepare Percoll gradients in 15 ml conical tubes (1 tube prep for 10 pups, so 8 gradients are needed for 80 pups). Add 3 ml "Bottom" Percoll solution to each tube. Gently and slowly pipette 3 ml "Top" Percoll solution, taking care to maintain separation of the solutions (if desired, phenol red can be added to one of the solutions for better visibility).

10. After the final enzyme solution incubation, use a 70 µm sieve to filter the combined cell suspension, centrifuge at $700 \times g$ for 10 min, and resuspend pellet in 16 ml of dispersion media (2 ml/10 pups). Carefully layer onto Percoll gradients, taking care not to mix with the existing Percoll solutions.

11. Centrifuge the gradients at $2,060 \times g$ for 27 min; do not use a centrifuge break (30 min total spin).

12. Collect fibroblasts and myocytes into separate 50 ml conical tubes. First, aspirate the top pink media layer. Immediately beneath this (~5 ml mark on the conical) is the fibroblast layer; add to a 50 ml conical containing fibroblast growth medium (10 % FBS in DMEM with 1 % L-glutamine, 1× NEAA, and 1× Pen/Strep). The myocytes are contained in the next clear buffy layer (~3 ml mark on the conical). Collect just above and below this layer for optimal results; collecting too much more than this, however, will lead to stem cell and endothelial cell contamination in cultures. Add collected myocytes to a 50 ml conical tube containing 5 ml of dispersion medium.

13. Centrifuge collected cells at $700 \times g$ for 10 min, and count cells. Fibroblasts can be plated on five 0.1 % gelatin-coated T-175 flasks; 1 h later, rinse the plates of debris and refeed with fibroblast growth medium.

14. Plate myocytes according to the desired density in dispersion media. They can be grown on 0.1 % gelatin, fibronectin, or collagen; the substrate on which they are plated will affect their survival and phenotype, and should be selected carefully. Recommended plate densities for a confluent culture are as

follows: 6-well plates, 1.25×10^6 cells/well; 35 mm adhesion dishes, 1×10^6 cells/well; 24-well plate, 6×10^4 cells/well; 48-well plate: 4×10^4 cells/well.

15. Feed myocytes approximately 24 h after plating. Within 24–48 h, myocytes should start beating and will eventually be synchronous within the well; cells should not have many vacuoles or too much membrane blebbing. Cells should initially be grown in dispersion medium (10 % horse serum/5 % FBS); serum-free medium may be used as needed for experiments, but it should be noted that myocytes begin to die within 24 h of serum removal. A lower amount of serum (1–5 %) is recommended to improve survival.

3.3 3D Collagen Gel Assay

1. Prepare collagen gel matrix in a precooled 50 ml conical, constantly keeping on ice. Use caution when pipetting the collagen stock; its viscosity can alter the gel volume easily. Also be sure to *thoroughly* mix the gel without introducing a large amount of bubbles (*see* **Note 25**).

2. Trypsinize, wash, and spin down ECs and/or fibroblasts (or other cells of interest) and resuspend cell pellets in ice-cold 1× M199 at a concentration of 1×10^7 cells/ml. Resuspend the cells with a pipette to break up any clumps. Place cells on ice until use.

3. If several experimental groups are needed, separate collagen gel matrix into precooled 1.5 ml microcentrifuge tubes. A preparation of 500 µl should yield 16–18 gels.

4. Add a total of 200 µl of cold cell suspension to 1 ml of the cold collagen gel matrix for a final concentration of 2×10^6 cells/ml. For cocultures, add 100 µl of ECs + 100 µl of fibroblasts (or other cells of interest) to the collagen gel matrix. Gently but thoroughly pipette to mix with collagen solution without making any air bubbles.

5. Add the cell–collagen gel mix at 28 µl per well in a 96 half-area well clear flat-bottom TC-treated microplate. After every third to fourth well, tap the plate gently on each side to evenly spread out the gel matrix within each well.

6. Place the plate in an incubator (37 °C with 5 % CO_2) for 30 min to allow collagen to polymerize and equilibrate.

7. Add 100 µl of feeding medium to each well, with or without growth factors or other molecules of interest.

8. Incubate the plate in 37 °C with 5 % CO_2 and allow ECs to undergo lumen and tube formation. Without growth factors added, 24 h is an optimal time point to quantify lumen and tube formation (Fig. 1).

9. If longer culture times are desired, replace 60 µl of the feeding medium after 2 days in culture. Complete replacement of media is not recommended.

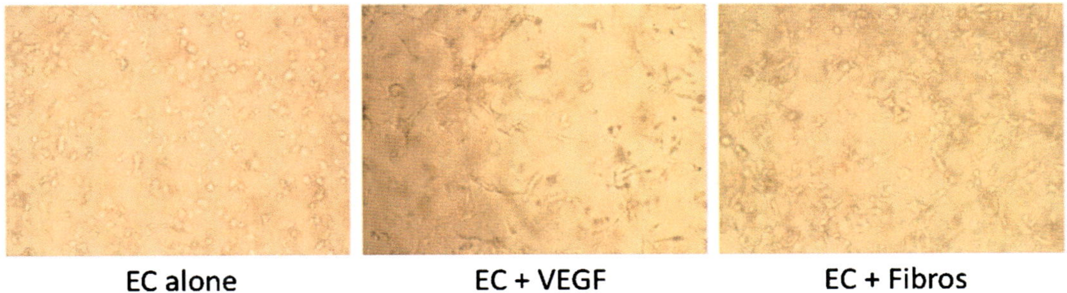

Fig. 1 3D collagen tube formation with cardiac fibroblasts and endothelial cells. Cardiac fibroblasts and endothelial cells were isolated as described, plated in collagen gel, and allowed to incubate in culture for 24 h. Endothelial cells were plated alone (*left panel*), with VEGF (*middle panel*), or with cardiac fibroblasts (*right panel*). Endothelial cells plated alone show little tube formation; however, endothelial cells plated with VEGF show numerous tubes, as do endothelial cells that have been cocultured with fibroblasts

10. When the desired endpoint is reached, aspirate media, gently wash plates with PBS, and then add the desired fixative. 2–4 % fresh paraformaldehyde can be used for common immunofluorescent stains and/or cryosectioning. 3 % glutaraldehyde is required for 0.1 % toluidine blue stain for visualization of cultures. Gels should be fixed 1 h to overnight at 4 °C.

3.4 3D Cell Adhesion Assay

1. Plate 1×10^6 freshly isolated myocytes on aligned collagen-coated 35 mm dishes. Collagen is aligned by tilting the plate during the collagen polymerization process. Additional layers of freshly isolated myocytes can be added on top of the original myocyte layer at 4-day intervals, resulting in multiple layers of myocytes.

2. After 48 h, when the myocytes have established their rod shape and are beating, add 5×10^5 neonatal cardiac fibroblasts. An additional layer of collagen may be added to the final layer of cells so that the entire culture is encased in aligned collagen. For the majority of cell–cell interaction studies, a single layer of myocytes is plated onto aligned collagen, and then 3 days later cardiac fibroblasts are plated on top of the myocytes.

3. Adhesion between fibroblasts and myocytes may be assayed as follows: Plate myocytes with cardiac fibroblasts on aligned collagen (**steps 1–2**) in the absence or the presence of various blocking antibodies or reagents to examine their effect on cell–cell interactions. Individual dishes should be plated for each of the time points desired so that the cells will be subjected to minimal agitation. At various time intervals (4, 8, 12, and 24 h after adding fibroblasts), gently swirl the plate, collect 50 μl of medium, and count the viable cells using a hemocytometer or a cell counter. The number of cells collected should be extrapolated for the total volume of media, and represent the number of fibroblasts no longer attached or adhered to myocytes (Fig. 2).

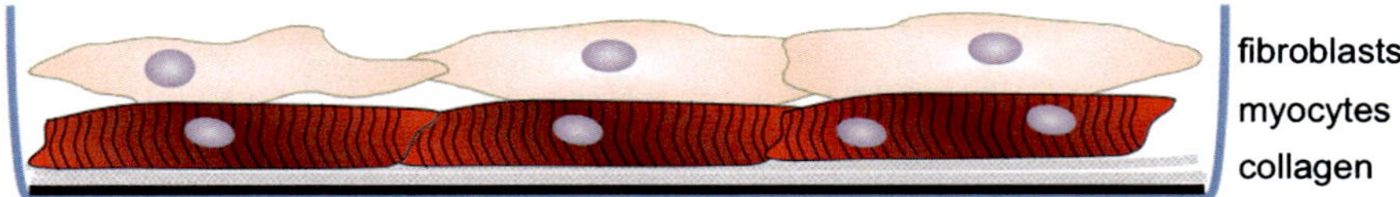

Fig. 2 Diagram of 3-dimensional cell culture system. Myocytes are plated onto aligned collagen and allowed to adhere and begin beating (24–48 h). At this point, fibroblasts are plated on top of the myocytes in the absence or the presence of blocking antibodies or other reagents to exam the effects on cell–cell interactions

3.5 Cell Aggregation Assay

1. Coat 50 ml Erlenmeyer flasks with 10 ml 3 % BSA/PBS overnight (on a platform rotator in the incubator), and then rinse with desired feeding media for 30 min prior to beginning the assay. The number of flasks depends on the number of experimental groups being tested.

2. Add 3×10^6 freshly isolated myocytes and 3×10^6 neonatal fibroblasts in 10 ml of desired media, and subject to rotational culture using a platform shaker at 80 rpm at 37 °C, 5 % CO_2 for 24 h. Myocyte-alone and fibroblast-alone control cultures should also be used. Individual aggregates are isolated using a dissecting microscope (*see* **Note 26**).

3. Aggregates can be disrupted into single-cell suspensions (via pipette) to count cell populations within each aggregate. In addition, whole aggregates may be fixed in fresh 2 % paraformaldehyde for further immunohistochemical analyses.

4. For studies involving disruption of cell–cell contacts, pre-incubate fibroblasts with the antibody or the reagent of choice for 30 min prior to coculture with myocytes.

4   Notes

1. When using laminin, coat the plates immediately before plating, without letting them dry.

2. Batch-to-batch variability of collagenase exists; adjust accordingly.

3. Add 1 % bovine serum albumin (BSA) to the DPBS if too much clumping occurs.

4. 15 ml conical tubes may also be used with a larger magnetic tube holder—adjust accordingly for users' optimal isolation.

5. Other Dynabeads are available to complement nearly any species of primary antibody generation.

6. DDR2 antibodies can be problematic, and differential plating may also be used to isolate fibroblasts. To do so, instead of incubating a second time with conjugated antibodies, substitute Subheading 3.1.4, **step 8**, by plating the flow through

after the CD31 incubation on a T-75 flask for approximately 1 h, and rinse several times with cell isolation medium (even tapping the plate to remove loose cells) before feeding with fibroblast growth medium.

7. Low-concentration gels are used for the described studies. It is recommended to test the cells and their optimal concentration prior to beginning an experiment. Note: 1 ml is a standard volume; 500 µl may be prepped, but because of the viscosity and pH, smaller preparations are not recommended.

8. This experiment cannot be scaled up to a full 96-well plate or larger.

9. Depending on the experimental parameters, 1 % FBS/DMEM (or serum-deprived media) may be used. This would also serve as a negative control.

10. See the Dynabead package insert for specific buffer to be used.

11. Overnight incubation provides optimal results, but 2–4 h at room temperature is acceptable.

12. According to the manufacturer, and the Dynabead package insert, the beads may be rewashed after this period, but antibody loss will greatly affect the yield of the isolation. It also notes that sodium azide may be used as a preservative and then washed out prior to use, but if the conjugated beads are used within 1–2 weeks this is not necessary.

13. Do *not* include the aorta in the digestion, and be consistent about where the cut is made. If ventricular cells are desired, quickly cut away the atria, maintaining consistency with each dissection.

14. No antibiotics are necessary, but the dissection must be performed quickly and the dish taken to the sterile hood as soon as possible. If time is a concern, antibiotics should be added to the cold cell isolation medium.

15. This step must be done quickly, but inefficient mincing will result in poor digestion and yield. Take care not to leave any large chunks.

16. This is a critical step in the digestion. Be sure to forcefully triturate the cells against the corner edge of the bottom of the conical, without splashing media; the first few times, some tissue will still be too large to fit into the cannula, but most should be readily aspirated. If trituration is incomplete, a large amount of cell clumping will appear in cultures.

17. If a good digestion is not achieved, increase the collagenase concentration next time, or perform a more thorough trituration, taking care to forcefully push the cells against the bottom/side of the tube.

18. Gey's solution may also be used to lyse red blood cells, if they are not adequately cleared during dissection. Be sure to centrifuge the cells again, and resuspend in 1 ml of DPBS for antibody incubation.

19. After the procedure is performed consistently, this is not entirely necessary; cell yield at this point and cell growth following plating do not always correlate.

20. Incubation may be increased up to 20 min, but a longer incubation will result in a negative sort (see Dynabead insert manual for additional details).

21. If it is suspected that the ECs may have altered adhesion properties (i.e., from a knockout mouse or an injury model), wait 36–48 h to rinse.

22. When becoming accustomed to the isolation technique, it is a good idea to plate the flow through on a T-75 flask (even when not isolating a second cell type) and culture for 2 days; these cells can also be sorted with a second sort (*see* Subheading 3.5).

23. If yield is good, and few contaminating cells were present, the culture can be plated for a 1:2 split at this time.

24. It is important to keep cells fairly confluent to avoid senescence. For best results, do not use cells cultured beyond passage 3.

25. It is often easier to use a larger (1,000 μl) pipette to mix. If low gel volume is a recurring problem during preparation and too much is being lost in the pipette tip, first measure the desired amount of sterile water, draw a line on the tube at the location of the water meniscus, aspirate the water, and add collagen to the marked line. This may help save time and prevent the collagen from polymerizing before desired.

26. We generally use myocytes, fibroblasts, and endothelial cells in our studies, but you can replace them with your choice of different cell types. We have previously used cancer cells and fibroblasts in these studies.

Acknowledgments

We acknowledge support from the American Heart Association (13GRNT14680067, T.A.B.).

References

1. Chang HY, Chi JT, Dudoit S et al (2002) Diversity, topographic differentiation, and positional memory in human fibroblasts. Proc Natl Acad Sci USA 99:12877–12882

2. Chipev CC, Simon M (2002) Phenotypic differences between dermal fibroblasts from different body sites determine their responses to tension and TGFβ1. BMC Dermatol 2: 1–13

3. Baudino TA, McFadden A, Fix C et al (2008) Cell patterning: interaction of cardiac myocytes and fibroblasts in three-dimensional culture. Microsc Microanal 14:117–125

4. Koh W, Stratman AN, Sacharidou A et al (2008) In vitro three dimensional collagen matrix models of endothelial lumen formation during vasculogenesis and angiogenesis. Meth Enzymol 443:83–101

Production of Spontaneously Beating Neonatal Rat Heart Tissue for Calcium and Contractile Studies

Fnu Gerilechaogetu, Hao Feng, Honey B. Golden, Damir Nizamutdinov, Donald M. Foster, Shannon Glaser, and David E. Dostal

Abstract

Neonatal rat ventricular myocytes (NRVM) and fibroblasts (FB) serve as in vitro models for studying fundamental mechanisms underlying cardiac pathologies, as well as identifying potential therapeutic targets. Typically, these cell types are separated using Percoll density gradient procedures. Cells located between the Percoll bands (interband cells [IBCs]), which contain less mature NRVM and a variety of non-myocytes, including coronary vascular smooth muscle cells and endothelial cells (ECs), are routinely discarded. However, we have demonstrated that IBCs readily attach to extracellular matrix-coated coverslips, plastic culture dishes, and deformable membranes to form a 2-dimensional cardiac tissue layer which quickly develops spontaneous contraction within 24 h, providing a robust coculture model for the study of cell-to-cell signaling and contractile studies. Below, we describe methods that provide good cell yield and viability of IBCs during isolation of NRVM and FB obtained from 0- to 3-day-old neonatal rat pups. Basic characterization of IBCs and methods for use in intracellular calcium and contractile experiments are also presented. This method maximizes the use of cells obtained from neonatal rat hearts.

Key words Cardiac myocytes, Fibroblasts, Neonatal, Density separation, Interband cells

1   Introduction

Primary cultures of neonatal rat ventricular myocytes (NRVM) and fibroblasts (FB) are widely accepted as in vitro models in basic cardiac research. These cells readily attach to cell culture surfaces and serve as a means to study a variety of pathophysiologic processes, which lack the influences of hemodynamic factors existing in vivo. The use of NRVM and FBs requires that both cell types be isolated as primary cultures. Successful isolation of these cell types from neonatal rat hearts requires that cells be dissociated from the heart tissue and purified. We have found that cocultures of the remaining cardiac cell types, termed interband cells (IBCs) due to location in the Percoll gradient during NRVM and FB purification, readily

Troy A. Baudino (ed.), *Cell-Cell Interactions: Methods and Protocols*, Methods in Molecular Biology, vol. 1066, DOI 10.1007/978-1-62703-604-7_4, © Springer Science+Business Media New York 2013

create a well-established coculture system that does not require external electrical stimulation to maintain rapid synchronous beating while in culture. The IBC monolayer has higher synchronous beating after 24–48 h and can be placed at higher rates (~180 bpm) compared to pure cultures of NRVM (90–120 bpm). The well-developed automaticity in the IBC monolayer suggests that it may serve as a new in vitro model for the study of cell-to-cell signaling and regulation of contractile responses. Cardiac myocytes in the IBC fraction readily attach to ECM-coated glass and plastic coverslips and culture dishes, which may be related to the more juvenile state of the myocytes upon plating. In the procedures below, we describe the isolation of IBCs as a by-product of NRVM and FB, as well as demonstrate some of the contractile properties of these cells. The cardiac cell isolation procedure described below yields approximately 1×10^6 NRVM and 1.5×10^6 IBCs from 40 animals, with 90 % viable cells. The complete procedure takes 8 h.

2 Materials

2.1 Equipment

1. Digital heating circulating water bath, 6 L volume.
2. Digital stir-plate with LED speed display.
3. Hemocytometer.
4. Horizontal laminar flow hood.
5. Jacketed double-sidearm cell spinner flask.
6. Monofilament open mesh (200 μm) filter material.
7. Refrigerated centrifuge (equipped with swinging bucket rotor).
8. Slide warmer.
9. Tissue culture microscope equipped with 40× fluorescent objective (UApo/340 nm, NA = 0.9, Olympus, Center Valley, PA, USA), excitation and emission filters for detection of fura-2.
10. Fotonic Sensor (MTI 2100, MTI Co., Latham, NY, USA).

2.2 Surgical Instruments

1. Fine "needle-nose" forceps.
2. Fine surgical scissors.
3. Gem single-edge razor blades.
4. Small surgical scissors.
5. Small surgical forceps.

2.3 Tissue Culture Plasticware

1. 10 ml serological transfer pipettes.
2. 15 and 50 ml sterile conical centrifuge tubes.
3. 500 ml (0.45 μm pore size) polyethersulfone bottle top filters.

    4. T-75 cell culture flasks.

    5. 100 mm×20 mm tissue culture dishes.

2.4 Media and Reagents for Cardiac Cell Dispersion

1. Cell dispersion medium: Prepare by combining Dulbecco's modified Eagle's medium (DMEM) and Medium 199 (M199) in a 4:1 ratio; supplement with 10 % horse serum, 5 % fetal bovine serum (FBS), and 100 units/mL penicillin, 100 µg/mL streptomycin; pH to 7.4; and filter sterilize.

2. Hank's–HEPES solution: Prepare by adding 6 g of HEPES per liter of calcium- and magnesium-free Hank's balanced salt solution (HBSS; 25 µM final HEPES concentration), pH to 7.4, and filter sterilize.

3. Enzyme solution: Prepare by adding 750 U of bovine trypsin, 825 U of α-chymotrypsin, and 150 U of collagenase type II to 300 ml of Hank's–HEPES solution. The enzyme solution should be prepared just prior to use and filter sterilized (*see* **Notes 1** and **2**).

4. Percoll density gradient buffer (12.5×): Prepare by combining 8.47 g of NaCl, 5.96 g of HEPES, 1.24 g of glucose, 0.5 g of KCl, 0.17 g of NaH_2PO_4, and 0.25 g of $MgSO_4$ in a total volume of 100 ml of deionized water; pH to 7.4; and filter sterilize.

---

3   Methods

The dispersion and purification methods described below have been optimized for the isolation of NRVM, FB, and IBCs from 40 Sprague Dawley neonatal rat pups, which are 0–3 days of age.

3.1 Preparation for Dispersion of Cardiac Cells

1. Warm the dispersion medium, Hank's–HEPES, and enzyme solutions to 37 °C by placing on the slide warmer, located in the laminar flow hood (Fig. 1).

2. Connect the water jacket of the cell stirrer to the ports of the circulating water bath and ensure that the temperature of the running water is 37 °C (*see* **Note 3**).

3. Place a 150 mm sterile tissue culture dish in the hood for euthanizing neonate rat pups along with sterilized surgical instruments.

4. Place two 100 mm tissue culture dishes in the hood, and add 15 ml of warm, sterilized Hank's–HEPES solution to one dish and 5 ml to the other dish (*see* **Note 4**).

3.2 Removal and Mincing of Neonatal Rat Heart Ventricles

1. Following euthanasia, open the chest of the animal with small scissors and remove the heart with the small forceps.

2. To remove blood from the heart, transfer it to a 100 mm dish containing 15 ml of warm Hank's–HEPES solution.

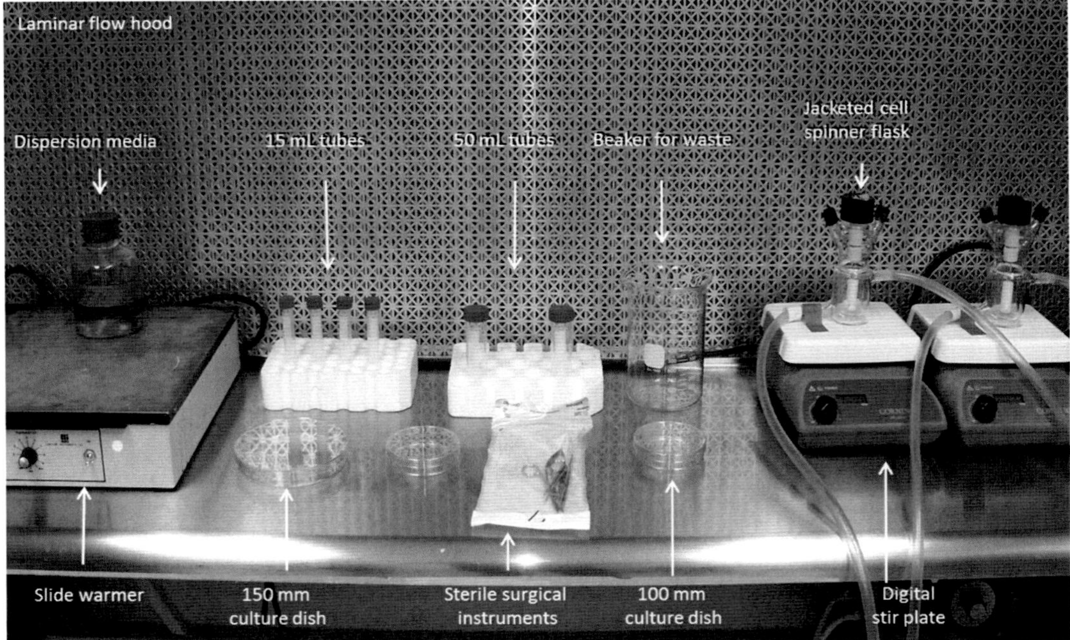

Fig. 1 Preparation for neonatal rat cardiac cell dispersion. Items required for harvesting neonatal rat heart ventricles, dissociation of cardiac tissue, and density gradient separation of cardiac cells are shown

3. Use the surgical forceps to transfer the heart to a second 100 mm dish containing 5 ml of warm Hank's–HEPES solution and remove the atria using the small surgical scissors.

4. Slice the heart into 1 mm pieces with the scissors.

5. Once all 40 hearts have been processed, use a sterile single-edge razor blade to comb the minced tissue to one side of the 100 mm culture dish. Use the razor blade to mince the ventricles into 0.5 mm pieces.

6. Gently transfer the mincing buffer (5 ml) and tissue pieces to a 50 ml water-jacketed tissue stirrer with a 10 ml pipette.

7. Add 7 ml of Hank's–HEPES solution to wash the 100 mm dish and transfer the solution and any remaining tissue pieces to the cell stirrer.

8. Repeat the above step with two 6 ml aliquots of warm enzyme solution. The total volume of the minced cardiac tissue in the cell stirrer should be 24 ml (*see* **Note 5**).

3.3 Enzymatic Dissociation of Cardiac Tissue

Cells in the myocardium are attached by weak cell-to-cell and cell-to-matrix interactions, which together are effective in retaining cells to the ventricular tissue. Therefore an enzymatic approach is required to dissociate NRVM, FB, and IBCs from the minced pieces of cardiac tissue. We use a combination of bovine trypsin, collagenase II, and α-chymotrypsin to break adhesions among the cells and digest extracellular matrix. Trypsin has been widely used as a dissociating agent for neonatal and adult rat hearts since its initial use in NRVM isolation procedures. This highly selective serine protease is used to degrade the peptide bonds in connective proteins, which increase susceptibility of the tissue to shearing forces during the dispersion. Collagenase II, which has greater clostripain activity, is used to dissociate the collagen fibers in the myocardium. The serine protease α-chymotrypsin is used to cleave cell surface proteins such as selectins and integrins, which attach cells to one another or extracellular matrix. Although cruder enzyme preparations are typically more effective due to the presence of other proteases, polysaccharidases, and lipases, purer preparations tend to be less cytotoxic. Below are detailed procedures for enzymatic dissociation of cardiac cells from neonatal rat hearts.

1. Secure the cell stirrer lid and adjust the stirring speed to 300 rpm.

2. After 12 min, turn off the stirrer and allow the tissue pieces to settle.

3. Gently remove the wash solution using a sterile 10 ml pipette. The pipette tip should be placed at the surface of the enzyme solution to prevent disruption of the tissue pieces. Discard the wash solution into a 1 L beaker.

4. Add 30 ml of fresh enzyme solution to the stirrer and digest the cardiac tissue for 15 min with the stirrer set at 600 rpm.

5. After 15 min, stop the stirrer and gently transfer the enzyme solution into a 50 ml centrifuge tube containing 10 ml of 25 °C dispersion medium. Using the pipetting technique described in **step 3**, be careful not to transfer tissue pieces or partially dissociated cell clumps.

6. Sediment the collected cardiac cells by centrifugation at $900 \times g$ for 10 min at room temperature.

7. Carefully decant the supernatant and discard into a 1 L beaker.

8. Gently resuspend the cell pellet in 5 ml of dispersion medium and collect in a 50 ml tube.

9. Repeat **steps 4–8** for an additional 4–6 digestions to remove the remaining cardiac cells from the tissue fragments. Following the final digestion, only extracellular matrix and DNA from damaged cells should remain. The remaining tissue pieces should be white and the cell solution clear.

10. Gently resuspend any cell clumps among the pooled cells by pipetting, and add enough dispersion medium for a final volume of 35 ml.

11. Pass resuspended cells through 200 μm Nitex mesh fabric to remove cell debris, and collect the cells in a sterile 50 ml tube.

12. Sediment the cells at $900 \times g$ for 10 min at room temperature and resuspend collected cells in 8 ml of dispersion medium.

Table 1
Preparation of discontinuous Percoll gradients

Number of pups	Number of gradient tubes	Volume of resuspended cells (ml)	Layer	12.5× Density buffer (ml)	Sterile water (ml)	Percoll (ml)
40–50	4	8	Low density	1.04	6.64	5.32
			High density	1.04	4.02	7.94
60–70	6	12	Low density	1.56	9.96	7.98
			High density	1.56	6.03	11.91
80–90	8	16	Low density	2.08	13.28	10.64
			High density	2.08	8.04	15.88

3.4 Density Gradient Separation of NRVM, FB, and IBCs

In this procedure, a two-step discontinuous Percoll gradient is used to isolate IBCs, as well as highly pure populations of NRVM and FB. The steps for preparing the Percoll solutions used to make the density gradient and separation of cardiac cells for 40 rat hearts are described in Subheading 2.4. Table 1 describes the preparation of these solutions for an additional number of rat hearts.

1. To each of the four 15 ml conical centrifuge tubes, create a two-layer density gradient by carefully layering 3 ml of the 1.060 g/ml Percoll solution on top of 3 ml of the 1.086 g/ml Percoll solution. The tube can be tilted to help minimize mixing of the Percoll layers during layering (Fig. 2).

2. Use a sterile 5 ml pipette to slowly add 2 ml of cell suspension (obtained in Subheading 3.3) to each of the tubes prepared above.

3. Centrifuge the tubes at $1,800 \times g$ for 45 min at room temperature.

4. After centrifugation, use sterile 5 ml transfer pipettes to harvest cells in the upper band cells (primarily FB), IBCs, and lower band cells (NRVM) (Fig. 2) into separate 50 ml tubes containing 15 ml of dispersion medium.

5. Repeat **step 4** for the remaining three tubes.

6. To each of the 50 ml tubes containing harvested cardiac cells, add dispersion media to make a final volume of 25 ml.

7. Sediment the harvested cells at $900 \times g$ for 10 min at 25 °C and discard the supernatant.

8. Add 5 ml of dispersion media and gently resuspend the cell pellets by pipetting.

9. Use the 5 ml pipette to measure the volume of resuspended FB. This should be carefully performed to prevent shearing of the cells, in case the pipette tip has sharp edges.

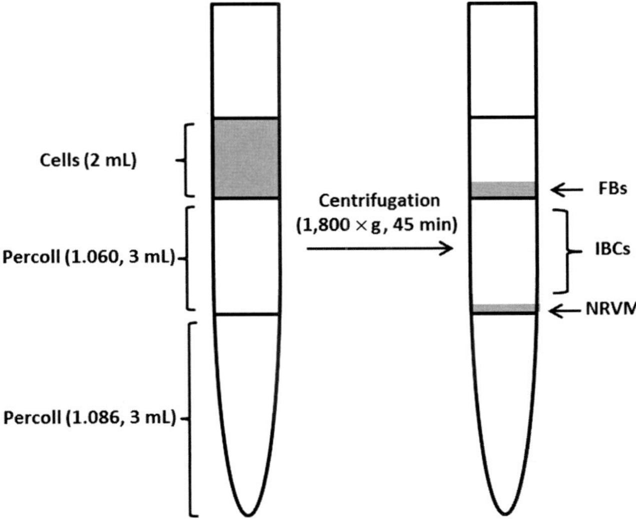

Fig. 2 Percoll gradient tube preparation and isolation of cardiac cells. The low-density Percoll solution (3 ml) is slowly layered over 3 ml of the high-density Percoll solution to form a discontinuous gradient. The cell suspension (2 ml) is then layered on the top of the low-density Percoll solution. After centrifugation, the FB band is located at the top of the low-density Percoll, whereas the NRVM are located at the interface of the low- and high-density gradients, and the IBCs are located between FB band and NRVM band

10. Repeat **step 9** for the tube containing NRVM and IBCs.

11. Add sufficient dispersion medium to the tubes containing NRVM, FB, and IBCs to final volumes of 30, 15, and 15 ml, respectively.

3.5 Harvesting FB (Upper Percoll Band)

The upper band of cells generated by the Percoll density gradient (described in Subheading 3.4) primarily contains FB. Contaminating cells, primarily endothelial cells, can be removed using differentially plating procedures. This procedure takes advantage of the ability of FB to more readily attach to the culture dish, compared to other cell types. Details of this procedure are as follows:

1. Equally divide the 24 ml volume of FB harvested in Subheading 3.4 into two T-75 culture flasks and incubate for 90 min at 37 °C in a 5 % CO_2 tissue culture incubator.

2. Remove the flasks from the incubator and firmly tap the bottom of the flask on the surface of the laminar flow hood to dislodge non-fibroblasts.

3. Remove the medium by aspiration, add 12 ml of warm dispersion medium to each of the flasks, and incubate cells for 24–48 h until confluent. At this point, FB can be passaged into experimental culture dishes.

4. To passage, wash FB cultures with warm phosphate buffer solution (PBS) and incubate with 3 ml of warm TrypLE (37 °C) per T-75 flask for 3–6 min. During this time, most of the cells should detach from the culture dish.

5. Transfer dissociated FB into 15 ml of dispersion media, count, and plate cells.

3.6 Harvesting IBCs (Cells Between Percoll Bands)

The IBCs are located between the upper and lower bands of the Percoll gradient and consist primarily of less mature NRVM, some FB, and vascular smooth muscle (VSM) cells. Prior to plating, it is important to determine the viability and number of IBCs by counting the number of cells that exclude trypan blue dye. Cells which exclude the blue stain are considered viable and used to calculate an index of viability. The above isolation procedures typically provide approximately 1.5×10^6 IBCs per 40 neonatal rat hearts with 70–90 % viability. These determinations are performed as described below:

1. Add 10 μl of the cell suspension to a 1.5 ml microcentrifuge tube containing 10 μl of trypan blue (0.4 %) and allow 1–2 min for dye absorption. Count both the total number of cells and the number of stained (dark) cells by a hemocytometer. Use the following equations to calculate the yield and viability:
   Yield = (total number of cells in four grids/4) × (5×10^4) × (volume cell suspension); percent viability = (total cells counted − stained cells)/total cells counted × 100.

2. Once the number of viable cells has been determined, calculate the final dilution volume for the harvested IBCs. Calculated volume = Yield/(desired cell density per plate) × (required volume per plate); final dilution volume = (calculated volume) − (volume cell suspension).

3. Plating density is 1×10^5 cells/coverslip.

3.7 Harvesting NRVM (Lower Percoll Band)

The lower band of cells generated by the Percoll density gradient (described in Subheading 3.4) primarily contains NRVM. The viability and number of NRVM are determined by counting the number of cells that exclude trypan blue dye, as described in Subheading 3.5. The above isolation and purification procedures typically provide approximately 1×10^6 NRVM per 40 neonatal rat hearts with 70–90 % viability. Prior to plating, add cytosine arabinoside (0.1 μM final concentration) to prevent cell division of non-myocytes.

3.8 Cardiac Cell Attachment

Within 1 day after plating, IBCs and NRVM should be attached and exhibit spontaneous beating, in which myocytes in the IBC fraction beat 1.5–2 times greater than those from the NRVM fraction. On day 3, NRVM form monolayers in which cells form cell-to-cell contacts as evidenced by pseudopodia, whereas the IBCs have well-developed junctions between myocytes and non-myocytes

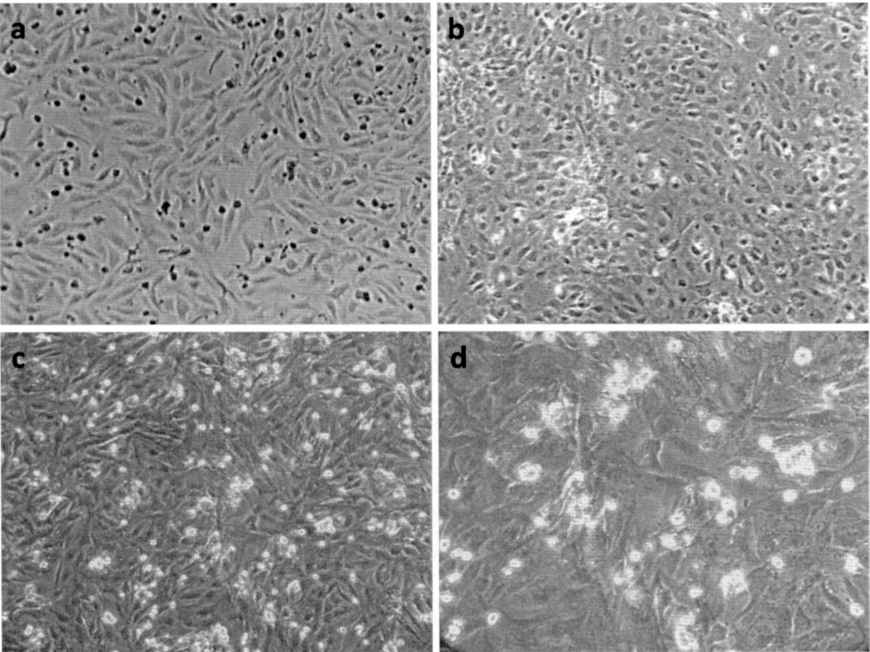

Fig. 3 Morphological features of NRVM, FB, and IBCs in culture. (**a**) NRVM plated on a collagen I-coated (1 µg/cm^2) 6-well plate, at a density of 0.8×10^6 cells/well, display cell spreading, but also maintain cell–cell contact. (**b**) FB after 2 days of culture, plated on a collagen I-coated (1 g/cm^2) 6-well plate at a density of 0.4×10^6 cells/well. (**c, d**) IBCs plated on a gelatin-coated (1 g/cm^2) 60 mm plate. Photographs taken at 10× (**a–c**) and 40× magnification

(Fig. 3c, d). Myocytes in the IBC monolayer should have an intrinsic rate of 60–90 beats per minute (Fig. 4), compared to 20–30 beats per minute for NRVM cultures. FBs, which are mononucleated, are characterized by a flat, irregular shape. In culture, FBs proliferate to form a confluent monolayer with indistinguishable cell–cell adhesion contacts.

3.9 Determination of Functional Changes in IBCs

IBCs are responsive to humoral stimuli such as isoproterenol (Fig. 4a) and can be paced up to 3 Hz (Fig. 4b). IBCs also readily attach to deformable membranes coated with extracellular matrix and are responsive to mechanical stretch in which intracellular calcium can be measured using fura-2 (Fig. 5a–c). Collective contractile responses can be readily determined using Fotonic sensing technology (Fig. 6a, b) and edge-detection (Fig. 6c). It is unclear as to why IBCs quickly obtain automaticity in the absence of pacemaker cells, when placed into culture. Although these questions remain to be addressed, cultured IBCs may be an appropriate model for the study of several kinds of heart disease, especially the neogenesis or regeneration of cardiac myocytes during recovery from heart disease.

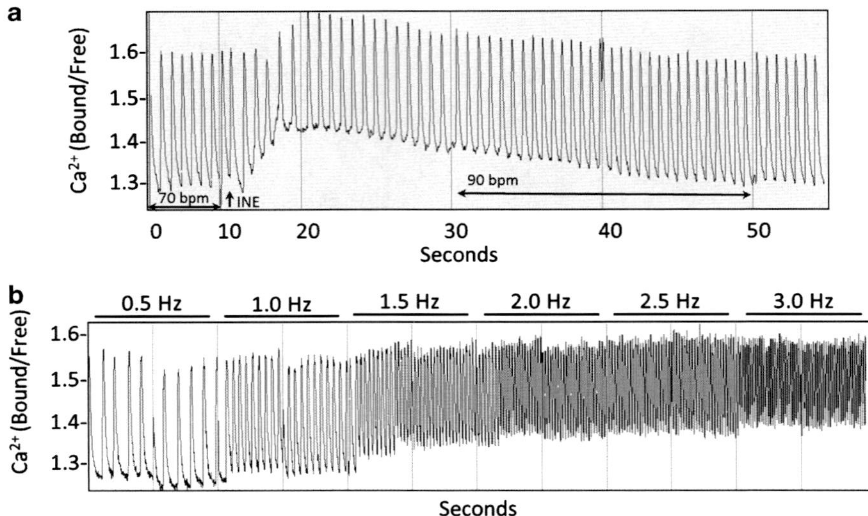

Fig. 4 IBCs spontaneously beat and respond to pharmacologic agents. IBCs were incubated with MEM containing 1 µM fura 2-AM (Molecular Probes) for 20 min at 37 °C and then washed with fresh MEM to remove excess dye. IBCs were imaged through a 40× Olympus objective and fluorescence measurements were recorded with a dual-excitation fluorescence photomultiplier system. Cells were exposed to light emitted by a 75 W lamp and passed through either a 360 or a 380 nm filter (±15 nm bandwidths) while being field stimulated to contract at 0.5–3.0 Hz (5 ms duration). Fura-2-loaded cells were excited at 360 ± 6.5 and 380 ± 6.5 nm with an ultraviolet xenon lamp. Emission fluorescence was measured at 510 ± 15 nm. Panel (**a**) demonstrates increased contraction force and intrinsic rate when IBCs were subjected to the beta-adrenergic agonist isoproterenol (10 µM). Panel (**b**) shows the frequency response of IBCs to electrical stimulation

4   Notes

1. Allow enzyme containers to reach room temperature prior to weighing. This will minimize moisture absorbed from the air.

2. Enzymes can substantially vary in efficiency of dissociation and toxicity among batches. Several lots of enzyme may need to be tested to obtain optimal results. Some vendors, such as Worthington Biochemical Company, provide a sampling program that can be used to choose the optimal batch of enzyme. Worthington Biochemical Company also maintains a Web-based database (http://www.worthington-biochem.com/cls/match.php), which can be used to match previous lots of the company's enzyme when reordering.

3. Laboratory tape attached to the bottom port of the jacket can be used to secure the cell stirrer to the stir plate.

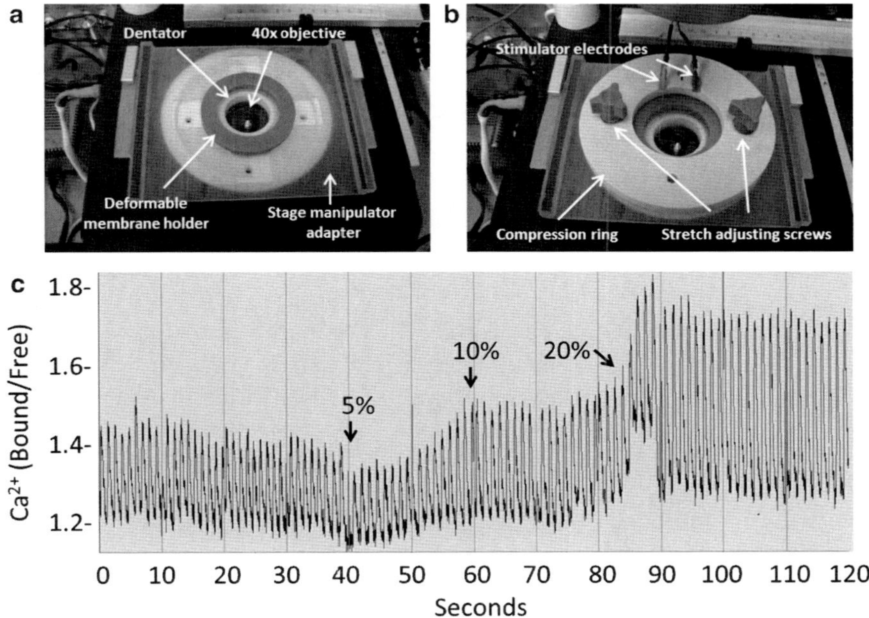

Fig. 5 IBCs have intracellular calcium responses to mechanical stretch. Panels (**a**, **b**) show the design of a customized stretch device that is used to visualize intracellular calcium changes upon mechanical stretch. The IBCs loaded with 1 µM fura-2 AM for 30 min were exposed to 5, 10, and 20 % of mechanical stretch (**c**). Cells were maintained in the focal plane of the objective during the stretch process. Cells were electrically stimulated at 1 Hz (7 V, 5-ms duration)

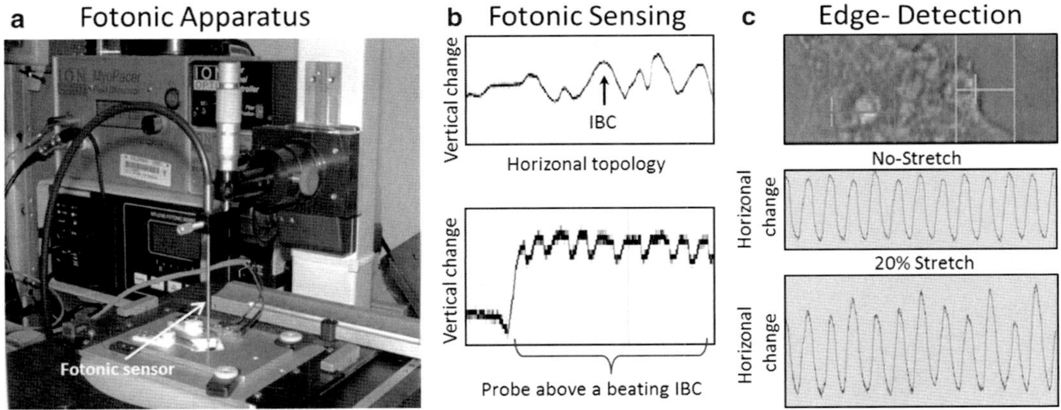

Fig. 6 Detection of contractile responses in IBCs using Fotonic Sensing and edge-detection. Changes in contraction resulting from IBCs can be determined using a Fotonic Sensor (**a**), which is a fiber-optic displacement measurement instrument equipped with a probe that detects vertical changes in the distance between the probe over an area of 150 µm (**b**), whereas horizontal changes can be detected using edge-detection (**c**). In this application, the probe was attached to the condenser arm of the microscope so that non-reflected light emitted by the probe could be used to visualize the field of cells undergoing contraction. Analog output from the Fotonic Sensor was directed to a preamplifier and analog digital converter and recorded using IonWizard software. Edge-detection and analyses were also performed using the IonWizard software

4. It is important to maintain the proper physiologic pH (7.4) for the buffers, enzyme solutions, and media used in the dispersion procedure.

5. Hearts must be isolated, minced, and transferred as quickly as possible to the enzyme solution to minimize cell death.

Acknowledgments

This manuscript is the result of work supported with resources and the use of facilities at the Central Texas Veterans Health Care System, Temple, Texas. Funding was provided the National Institutes of Health (5R01-HL068838-06), VA Merit Award (IO1 BX000801-01), VA hospital, and Scott and White Hospital.

Chapter 5

Paracrine Communication Between Mechanically Stretched Myocytes and Fibroblasts

Hao Feng, Fnu Gerilechaogetu, Honey B. Golden, Damir Nizamutdinov, Donald M. Foster, Shannon Glaser, and David E. Dostal

Abstract

Mechanical stretch is a major factor for myocardial hypertrophy and heart failure. Stretch activates mechanical sensors from cardiac myocytes, leading to a series of signal transduction cascades, which can result in cell malfunction and remodeling. It is well known that mechanical stretch also induces the release of paracrine factors from cardiac fibroblasts, as well as myocytes. Due to complicated circumstance of heart tissue, it is difficult to fully investigate the characteristics of these factors in situ. Here we describe static stretch and conditioned medium experiments as methods to examine the function of paracrine factors between primary cultured cardiac myocytes and fibroblasts.

Key words Cardiac myocytes, Cardiac fibroblasts, Neonatal, Density separation, Paracrine factors, Cardiac remodeling

1 Introduction

Mechanical stretch is a major factor towards the induction of myocardial hypertrophy and heart failure. These pathological responses can be partially modeled in vitro, where cardiac myocytes and fibroblasts are cultured on extracellular matrix (ECM)-coated flexible membranes and subjected to mechanical stretch [1, 2]. Mechanical stretch induces activation of signal transduction pathways, including mitogen-activated protein kinases (MAPK) cascades, phosphoinositide 3-kinase (PI3K)–protein kinase B (AKT) pathway, and phosphatases, that lead to abnormal myocyte function and hypertrophy [3]. A significant portion of the myocardium is composed of non-myocytes, primarily cardiac fibroblasts, that help maintain myocardial structure and function [4], and are activated in several cardiac pathologies including mechanical overload. It has been well established that cardiac fibroblasts produce factors that contribute to pathological cardiac hypertrophy characterized by increased cardiomyocyte size and reorganization of

Troy A. Baudino (ed.), *Cell-Cell Interactions: Methods and Protocols*, Methods in Molecular Biology, vol. 1066, DOI 10.1007/978-1-62703-604-7_5, © Springer Science+Business Media New York 2013

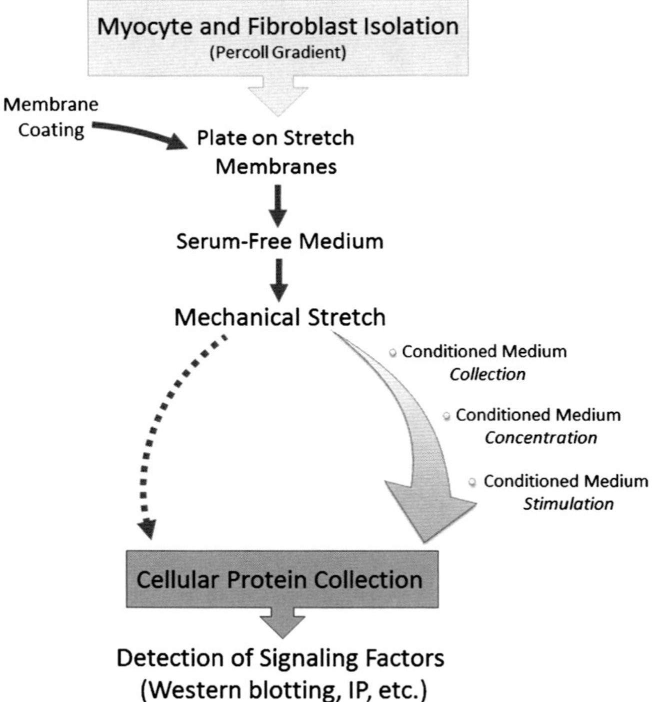

Fig. 1 Diagram of experimental protocol

contractile proteins. However, few studies have addressed the mechanisms by which stretched fibroblasts affect cardiac myocyte function. It is difficult to consistently evaluate the role of noncardiac paracrine factors in situ because the presence of multiple cell types (e.g., immune cells, endothelial cells), associated paracrine factors, ever-changing concentrations of neurohumoral factors, and the "washing" effect of surrounding capillaries creates a complex experimental environment. Therefore, in this book chapter, we describe the method of combining mechanical stretch and conditioned medium experiments to systematically assess the role of paracrine factors released from stretched cardiac fibroblasts on cardiac myocyte signaling (Fig. 1). Briefly, primary cultured cardiac myocytes and cardiac fibroblasts are subjected to static mechanical stretch. The medium conditioned by stretched cardiac fibroblasts is collected at different time points and applied to non-stretched cardiac myocytes, followed by protein sample collection and immunoblots. For the preparation of this methodology, we have used stretch-induced MAPK activation in neonatal rat cardiac myocytes as an example, which is directly activated by stretch and indirectly activated by paracrine factors secreted by cardiac fibroblasts. A similar mechanism may be involved in vivo in the mechanically stretched myocardium. While these studies are focused on the heart, these techniques can also be applied to other cell–cell communication models, such as with tumor cells and stromal cells.

2 Materials

Because cardiac fibroblasts and myocytes can have different signaling responses and release different factors, the success of these studies depends on the isolation of pure primary cultures of myocytes and fibroblasts. Our laboratory uses the Percoll gradient technique to obtain highly purified cardiac myocytes and fibroblasts from neonatal rat hearts. A detailed description regarding the isolation of cardiac myocyte and fibroblast cultures was recently described by our lab [5]. This technique is also presented in this current book (Gerilechaogetu et al.). All solutions are prepared from ultrapure water and analytical grade reagents.

2.1 Cell Culture Supplies

1. 10 ml serological transfer pipettes.
2. 15 and 50 ml sterile conical centrifuge tubes.
3. 500 ml (0.45 μm pore size) polyethersulfone bottle top filters.
4. Calcium- and magnesium-free Hank's balanced salt solution (HBSS).

2.2 Stretch Membrane Coating

1. Class II biosafety cabinet equipped with UV light.
2. Tissue cell culture incubator.
3. BioFlex 6-well plates (Flexcell International Corp., Hillsborough, NC, USA).
4. Mouse collagen IV.
5. 0.05 N HCl.
6. Sterile phosphate-buffered saline (PBS), pH 7.4.

2.3 Mechanical Stretch and Conditioned Medium Collection

1. Dispersion medium: Combine Dulbecco's modified Eagle's medium (DMEM) and Medium 199 (M199) in a 4:1 ratio and supplement with 10 % horse serum, 5 % fetal bovine serum (FBS), and 34 μg/ml ampicillin; pH to 7.4; and filter sterilize.
2. Serum-free medium: Combine DMEM and M199 in a 4:1 ratio and supplement with 100 μM ascorbic acid, 1 mg/l transferrin, and 10 μg/l sodium selenite; pH to 7.4; and filter sterilize.
3. FX-3000T flexercell strain unit equipped with gaskets, flexercell baseplates, flexlink, system controller, and vacuum system (Flexcell International Corp.).
4. Primary cultured cardiac myocytes and fibroblasts.
5. 10× Cell lysis buffer (#9803, Cell Signaling Technology, Danvers, MA, USA) with Complete protease inhibitor cocktail (sc-29130, Santa Cruz BiotechnologyDallas, TX, USA) and phosphatase inhibitors (1 mM Na_3VO_4, 0.5 mM NaF).
6. Cell scrapers.

2.4 Purification and Concentration of Conditioned Medium (Optional)

1. Refrigerated centrifuge that will accommodate 1.5, 15, and 50 ml conical tubes.

2. Centriplus concentrators (Amicon Ultra Filter Units, EMD Millipore, Billerica, MA, USA).

2.5 Stimulation of Cells with Conditioned Medium

1. Serum-free medium.

2. Primary cultured cardiac myocytes and cardiac fibroblasts.

3. 10× Cell lysis buffer (Cell Signaling Technology) with Complete protease inhibitor cocktail (Santa Cruz) and phosphatase inhibitors.

4. Cell scrapers.

3 Methods

3.1 Coating Stretch Membranes

The Flexercell strain unit is used to mimic the stretch conditions in the heart. To perform the stretch experiments, cells must be cultured on flexible membranes, such as BioFlex-collagen IV-treated 6-well plates. These membranes can be reused after cleaning and re-coating (*see* **Notes 1–3**). We have tested different ECM proteins and confirmed that collagen IV is the best matrix for cardiac myocyte and fibroblast growth (Fig. 2; *see* **Note 4**).

1. Place membranes and lids (inverted) in tissue culture hood under UV light for at least 15 min. This step is not required for new sterilized plates fresh out of the package.

2. Preparation of membrane coating solution:

    (a) Transfer the stock collagen IV solution from –80 °C to 4 °C 1 day prior to coating the membrane.

    (b) Dilute 1 mg collagen IV in 500 ml 0.05 N HCl to obtain a 2 μg/ml coating solution. This will provide enough collagen IV to coat 40 six-well plates with 4 μg/well of collagen IV. Mix the collagen solution thoroughly with a 10 ml transfer pipet.

3. Add 2 ml of collagen IV solution to each well. The coating solution should cover the entire surface of the deformable membrane. Store coated dishes overnight at 4 °C prior to use. Plates may be stored for up to 2 weeks with coating solution.

4. Preparation of membranes for plating cells:

    (a) Transfer coated membranes from 4 °C to the tissue culture hood and expose to UV light for 15–20 min.

    (b) Transfer the plates to tissue culture incubator (37 °C) for at least 30 min prior to plating cells.

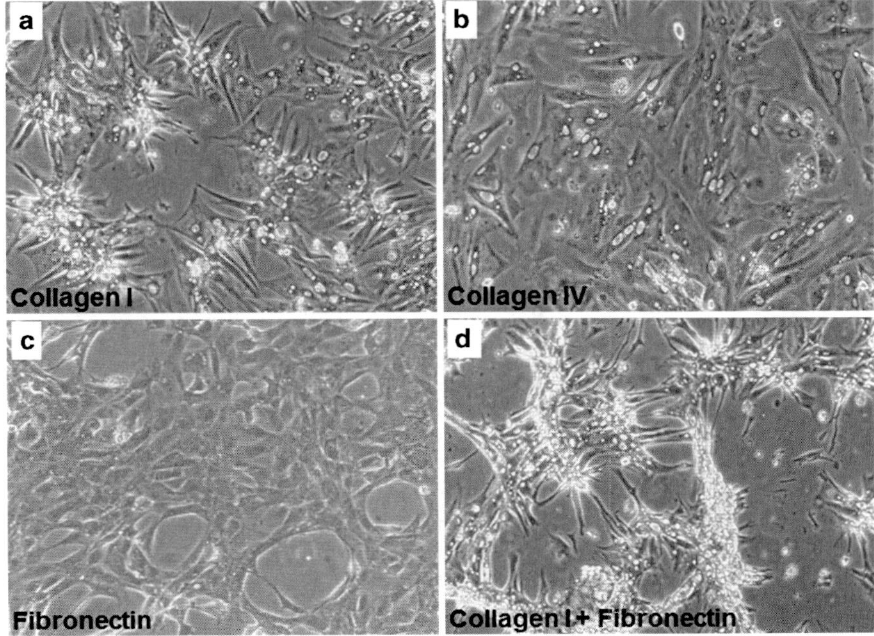

Fig. 2 Cardiac myocytes cultured in Bioflex 6-well plates coated with various types of extracellular matrix. NRVMs were plated on Bioflex 6-well plates coated with collagen I (**a**), collagen IV (**b**), fibronectin (**c**), or collagen I and fibronectin (**d**) at a density of 0.8×10^6 cells/well for 24 h. NRVMs plated on collagen I (**a**) or collagen I- and fibronectin (**d**)-coated plates show detachment and aggregation; cells plated on fibronetctin (**c**)-coated plates formed huge hole structure; cells on collagen IV (**b**) plates display mono layer spreading and maintaining cell–cell contact

(c) Aspirate the collagen IV coating solution from the wells and remove any condensation remaining on the inside of the lids.

(d) Add 2 ml of sterile PBS to each well and swirl plates to wash the walls of the wells.

(e) Aspirate the PBS from the wells and plate cells.

3.2 Cell Culture

Primary cultures of cardiac myocytes and cardiac fibroblasts were prepared from 0- to 3-day-old Sprague Dawley rats as previously described [5], as well as in this book (Gerilechaogetu et al.). For stretch experiments, both cardiac myocytes and cardiac fibroblasts are cultured on BioFlex-collagen IV-treated 6-well plates prepared as described above in Subheading 3.1 (Fig. 3).

3.3 Mechanical Stretch and Conditioned Medium Collection

The Flexercell strain unit is used to model in vivo mechanical stretch effects on cardiac myocytes and fibroblasts. Before starting the experiment, cardiac fibroblasts are growth arrested for 24 h in serum-free medium. Depending on the experimental conditions, static or cyclical stretch can be applied to the cells using the

NRVM   NRFBs

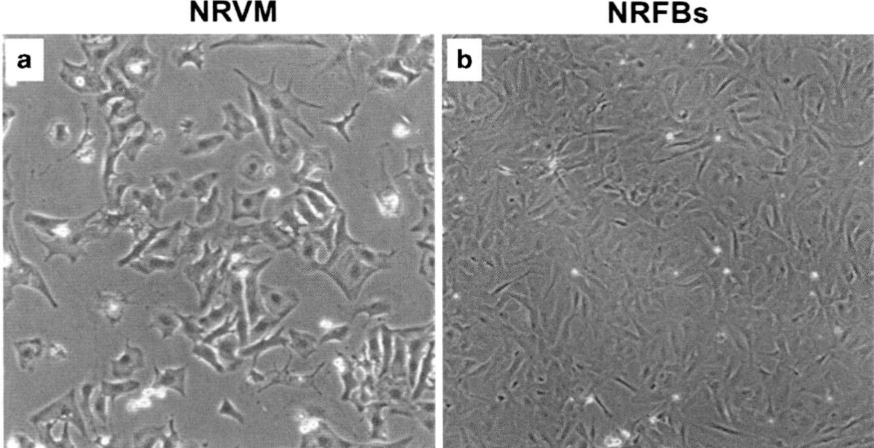

Fig. 3 Morphology of isolated cardiac cells. Percoll gradient methods were applied to get highly pure cardiac myocytes and cardiac fibroblasts. (**a**) NRVMs plated on a collagen IV-coated (4 μg/well) 6-well plate at a density of 0.8×10^6 cells/well display cell spreading but also maintain cell–cell contacts. (**b**) Fibroblasts after 2 days of culture plated on a collagen IV-coated (4 μg/well) 6-well plate at a density of 0.5×10^6 cells/well

computer-assisted Flexcell Tension system FX-3000T (Fig. 4). For the procedures described below, 20 % static stretch, which mimics the condition of cardiac hypertrophy [6], is applied for various periods of time, to obtain the temporal effects of paracrine factors on cellular signaling. This degree of stretch does not result in cell liftoff or decreased cell viability. The steps in stretching and harvesting the medium to examine paracrine effects are as follows:

1. Subject cultures of cardiac myocytes or fibroblasts to serum-free medium for 16–24 h prior to beginning stretch experiments (*see* **Note 5**).

2. Turn on the vacuum system and the FX-3000 Flexercell Strain Unit Controller.

3. Set up the FX-3000 Flexercell Strain Unit to 20 % stretch for 35 mm plates following the instruction manual (*see* **Note 6**).

4. Place the cells in the incubator of the FX-3000 Flexercell Strain Unit at least 40 min prior to stretching (*see* **Note 7**).

5. Stretch the cells for different periods of time, depending on your experimental design.

6. Reserve control plates, which do not undergo the stretch procedure.

7. At the end of each stretch period, remove plates from the manifold and place on ice and transfer media from the wells to 15 ml conical tubes. *This is the conditioned medium* (*see* **Note 8**).

8. To harvest cell extracts, add 50 μl ice-cold lysis buffer to each well, dislodge cells from membranes using a cell scraper, collect in 1.5 ml centrifuge tubes, and keep on ice (*see* **Note 9**).

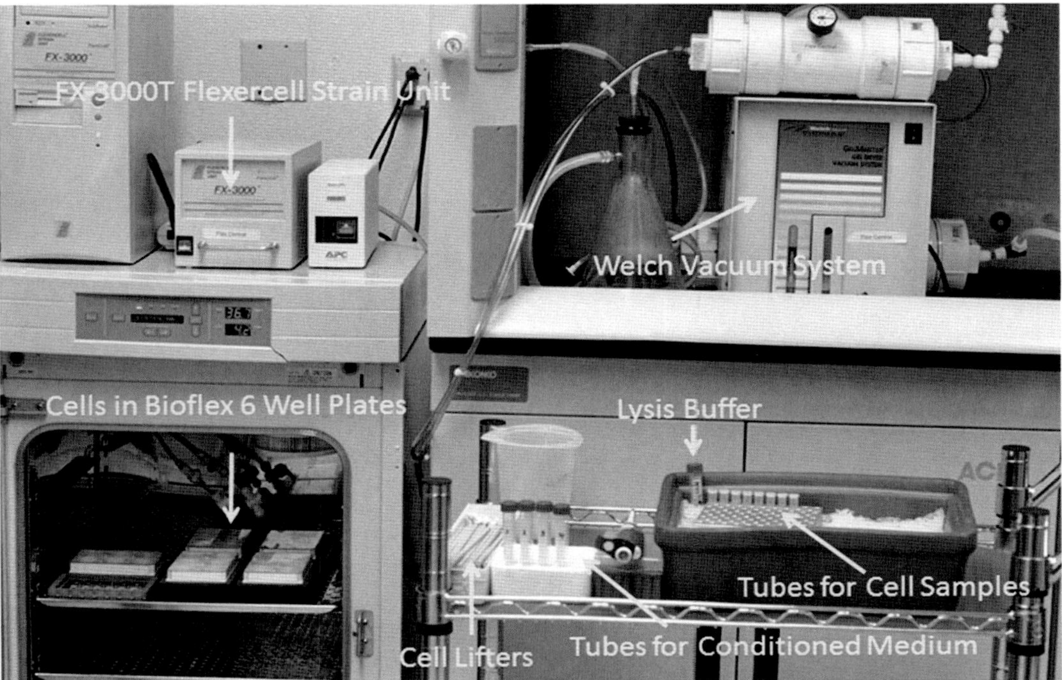

Fig. 4 Preparation for mechanical stretch. Items required for mechanical stretch, conditioned medium collection, and cell sample collection are shown

9. Vortex the cell lysates and sediment at $14,000 \times g$ for 10 min at 4 °C. Collect the supernatants and transfer to fresh 1.5 ml microcentrifuge tubes.

10. Store the cell extracts at −80 °C until use for protein determinations and Western blot analysis.

3.4 Purification and Concentration of Conditioned Medium (Optional)

In some cases, it may be necessary to fractionate and concentrate the conditioned media to identify the specific autocrine/paracrine factors involved in mediating signaling responses. Centriplus concentrators, which are available for various volumes of media and sizes of molecules can be used to achieve this purpose. Briefly, the procedure involves transfer of the conditioned medium to the concentrator's sample reservoir and the samples are fractionated according to the molecular weight cutoff of the device by centrifugation. Centrifugal force drives solvents and low-molecular-weight solutes through the membrane and into the filtrate vial. Retained macrosolutes remain above the membrane inside the sample reservoir. As the sample volume is diminished, retained solute concentration increases. In the recovery phase of the operation, sample is transferred to the retentate vial by placing the vial over the sample reservoir, inverting the device, and then centrifuging a second time. This centrifugal recovery method minimizes adsorptive losses to the membrane and reservoir wall. The specific steps in this procedure are as follows:

1. Insert the sample reservoir into the filtrate vial.

2. Add conditioned medium to the sample reservoir. Be careful not to touch the membrane with the pipette tip. Attach spin cap.

3. Place the reservoir assembly in the centrifuge rotor and counterbalance with a similar unit. Be sure that the rotor adapter can accommodate the entire device, as any obstruction could damage the concentrator during centrifugation.

4. Centrifuge at $3,000 \times g$ until the desired concentration is achieved (*see* **Note 10**).

5. Remove the concentrator from the centrifuge and separate the filtrate vial from the reservoir. For filtration applications, reserve the filtrate.

6. Place retentate vial over sample reservoir. Invert the assembly and centrifuge at $2,000 \times g$ for 3–4 min to transfer concentrate to retentate vial.

7. Remove device from the centrifuge. Separate the retentate vial from the reservoir and cover retentate vial with storage cap.

8. Concentrate and filtrate may now be stored at –80 °C for later use.

3.5 Testing Conditioned Medium for Signaling Effects

To test the paracrine effects of stretched cardiac fibroblasts, the conditioned medium collected from stretched cardiac fibroblasts is used to stimulate primary cultured cardiac myocytes and fibroblasts. The cultured cells should be in healthy condition and growth arrested for 16–24 h. Negative and positive controls are critical for any quantification of results. EGF stimulation is a good positive control for MAPK and AKT activation research. While it is difficult to determine the exact factor in the conditioned medium responsible for signaling, different pharmacological inhibitors, neutralizing antibodies, or siRNA can be used to find clues. ELISA or zymography may also be used to determine the target.

1. Subject the cardiac cells (myocytes or fibroblasts) to serum-free medium for 16–24 h before beginning experiment.

2. Pretreat naïve cells with serum-free medium for 40 min (*see* **Note 11**).

3. Change culture medium to conditioned medium obtained from stretch experiments.

4. Set up negative controls by adding unconditioned media to naïve cells.

5. After culturing with conditioned medium for experimental time points, aspirate or collect medium, add 50 µl ice-cold lysis buffer to each well, and collect samples the same way as described above (Subheading 3.3).

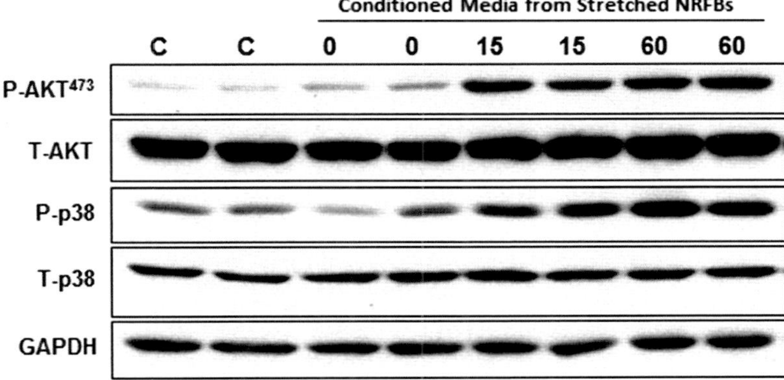

Fig. 5 Conditioned medium from stretched cardiac fibroblasts stimulates cardiac myocytes. Cardiac fibroblasts were stretched (20 %) for 15 or 60 min. After stretch, conditioned medium collected from different stretch time points and control plates was added to primary cultured cardiac myocytes. After stimulation for 15 min, cell samples from cardiac myocytes were harvested and analyzed by immnuoblotting with phospho-AKT and JNK antibodies, as indicated

3.6 Target Protein Detection

Depending on the experimental design, a variety of different protein detection methods (western blot analysis, co-immunoprecipitation, ELISA, etc.) can be used to detect target proteins. Here, we have used western blot analysis to reveal that conditioned medium from stretched cardiac fibroblasts significantly increases AKT and p38 phosphorylation in cardiac myocytes (Fig. 5).

4 Notes

1. In our experience, Bioflex 6-well plates can be reused at least 15 times after the wells are cleaned and following the coating procedures. The anionic detergent Alconox 1201 Liquinox is recommended for cleaning the membranes. Bleach should not be used since this cleaning agent will chemically modify the surface of the deformable membrane, thereby reducing attachment of ECM.

2. All the solutions for membrane coating must be sterilized.

3. Although Bioflex 6-well plates coated with collagen IV and other types of ECM can be purchased, non-coated plates allow for the surface coating to be customized according to individual lab's experimental conditions.

4. The collagen IV coating solution can be reused two times, with no significant effects on the attachment of cardiac myocytes or fibroblasts to the deformable membranes.

5. Cardiac fibroblasts can accommodate longer periods of time in serum-free medium compared to cardiac myocytes. We have

found that restricting serum starvation to 16 h substantially reduces the activation of signaling mechanisms related to cell death.

6. Insure that plates are properly seated into the gaskets, as inadequate stretch will result in the event of a vacuum leak around the plate.

7. It is recommended that conditioned medium stimulation experiments immediately follow stretch experiments. In this case, conditioned medium should be kept in the incubator to maintain proper temperature and pH.

8. Different lysis buffers can be chosen depending on the protein assay protocol used. For example, RIPA buffer is appropriate for co-immunoprecipitation experiments.

9. Do not exceed the maximum recommended force of $3,000 \times g$.

10. Serum-free medium and conditioned medium should be pre-warmed in a 37 °C water bath.

11. Incubation times are empirical and dependent on variables, such as the state of the cells and the effect times of any inhibitors used.

Acknowledgments

This manuscript is the result of work supported with resources and the use of facilities at the Central Texas Veterans Health Care System, Temple, Texas. Funding was provided the National Institutes of Health (5R01-HL068838-06), VA Merit Award (IO1 BX000801-01), VA hospital, and Scott and White Hospital.

References

1. Komuro I, Kaida T, Shibazaki Y et al (1990) Stretching cardiac myocytes stimulates protooncogene expression. J Biol Chem 265:3595–3598

2. Sadoshima J, Jahn L, Takahashi T et al (1992) Molecular characterization of the stretch-induced adaptation of cultured cardiac cells. An in vitro model of load-induced cardiac hypertrophy. J Biol Chem 267:10551–10560

3. Lal H, Verma SK, Golden HB et al (2008) Stretch-induced regulation of angiotensinogen gene expression in cardiac myocytes and fibroblasts: opposing roles of JNK1/2 and p38alpha MAP kinases. J Mol Cell Cardiol 45:770–778

4. Camelliti P, Borg TK, Kohl P (2005) Structural and functional characterisation of cardiac fibroblasts. Cardiovasc Res 65:40–51

5. Golden HB, Gollapudi D, Gerilechaogetu F et al (2012) Isolation of cardiac myocytes and fibroblasts from neonatal rat pups. Methods Mol Biol 843:205–214

6. Sadoshima J, Izumo S (1993) Mechanical stretch rapidly activates multiple signal transduction pathways in cardiac myocytes: potential involvement of an autocrine/paracrine mechanism. EMBO J 12:1681–1692

Chapter 6

Assessing Blood–Brain Barrier Function Using In Vitro Assays

Joseph Bressler, Katherine Clark, and Cliona O'Driscoll

Abstract

The impermeability of the blood–brain barrier (BBB) is due to a number of properties including tight junctions on adjoining endothelial cells, absence of pinocytic vesicles, and expression of multidrug transporters. Although the permeability of many chemicals can be predicted by their polarity, or oil/water partition coefficient, many lipophilic chemicals are not permeable because of multidrug transporters at the luminal and abluminal membranes. In contrast, many nutrients, which are usually polar, cross the BBB more readily than predicted by their oil/water partition coefficients due to the expression of specific nutrient transporters. In vitro models are being developed because rodent models are of low input and relatively expensive. Isolated brain microvessels and cell culture models each offers certain advantages and disadvantages. Isolated brain microvessels are useful in measuring multidrug drug transporters and tight junction integrity, whereas cell culture models allow the investigator to measure directional transport and can be genetically manipulated. In this chapter, we describe how to isolate large batches of brain microvessels from freshly slaughtered cows. The different steps in the isolation procedure include density gradient centrifugations and filtering. Purity is determined microscopically and by marker enzymes. Permeability is assessed by measuring the uptake of fluorescein-labeled dextran in an assay that has been optimized to have a large dynamic range and low inter-day variability. We also describe how to evaluate transendothelial cell electrical resistance and paracellular transport in cell culture models.

Key words Blood–brain barrier, Microvessels, Tight junctions, Transporters, Electrical resistance

1 Introduction

In vitro models of the blood–brain barrier (BBB) can be divided into two broad classes, cell culture and ex vivo. Cell culture models have been the more popular and are essentially based on a format that has been applied for modeling epithelial barriers [1–4]. Cells are plated on a semipermeable membrane on inserts that are fitted into multi-well plates. The two endpoints used to validate cell culture models are electrical resistance and drug permeability. Electrical resistance assesses tight junction integrity [5, 6] and in vivo, the electrical resistance of the BBB has been estimated to have an

Troy A. Baudino (ed.), *Cell-Cell Interactions: Methods and Protocols*, Methods in Molecular Biology, vol. 1066, DOI 10.1007/978-1-62703-604-7_6, © Springer Science+Business Media New York 2013

electrical resistance over $2,000\,\Omega/cm^2$ [7]. Several different in vitro models have reported electrical resistances between 700 and $1,450\,\Omega/cm^2$. These models differ in their level of complexity. One model uses endothelial cells differentiated from induced human embryonic stem cells cocultured with astrocytes [8] and the other uses endothelial cells isolated from porcine brain without astrocytes [4]. Still, a third model achieves an approximate resistance of $1,200\,\Omega/cm^2$ by applying a shear force to a monolayer of human brain endothelial cells by pumping buffer above the monolayer of cells. In this model, astrocytes do not have an effect on the electrical resistance [9]. The relation between the oil/partition coefficient and permeability across the endothelial cell monolayer in these in vitro models appears similar to the relation found in vivo [10–12].

To accurately predict drug transport, a model of the BBB must also mimic the kinetics of drug transporters. Currently, this is lacking in cell culture models. The ex vivo model described in this chapter are isolated brain microvessels (BMs), which are isolated from cow brains by homogenization, filtration, and sieving. In BM, the Vmax for p-glycoprotein was reported to be approximately tenfold greater than it is for brain endothelial cell cultures [13]. It is very likely that the kinetics for other drug transporters are also higher in BM. Another advantage of BM is that all of the known drug transporters are expressed. Many drugs are transported by more than one transporter, for example, the anthracyclines, daunorubicin, and doxorubin are transported by both p-glycoprotein and breast cancer resistance protein [14, 15]. One popular cell culture model for determining substrates for multidrug transporters are the Madin–Darby canine kidney (MDCK) epithelial cell line engineered to express different transporters [16, 17]. Their disadvantages are that (1) only one transporter is expressed; (2) it is not expressed at the level displayed by the BBB; and (3) it does not have the same posttranslational modifications as observed in brain endothelial cells.

Brain microvessels are fragments of capillaries, venules, and arterioles that retain the properties of the tight BBB [3, 18–20]. Chemicals that can destroy cells (e.g., trypsin) are not used in the isolation procedure and BMs are not cultured. BMs are predominantly composed of endothelial cells, while some pericytes and astroglial endfeet remain attached to the vessel wall. BMs maintain a very tight barrier; endothelial cells are adjoined by tight junctions, and neither fenestrae or pinocytic vesicles are observed [21]. They are impermeable to larger molecular weight chemicals, such as dextrans, and polar chemicals, such as sucrose [22]. A disadvantage of BM is that only the abluminal transporters face the buffer, so directional transport of chemicals across the BBB through luminal transporters (e.g., nutrients) cannot be assayed. Many drugs, however, are lipophilic and undergo paracellular transport or fail to cross because of the drug efflux transporters [23–27]. Consequently, directional transport is not a major concern.

2 Materials

2.1 Microvessel Preparation

1. Hank's Balanced Salt Solutions (HBSS).
2. HBSS supplemented with 1 % bovine serum albumin (BSA) (*see* **Note 1**).
3. HBSS supplemented with 30 % dextran (*see* **Note 2**).
4. 118 and 53 μm nylon mesh (Tetko, Elmsford, NY, USA). Soak the nylon mesh in HBBS/1 % BSA on the day that the brain microvessels will be prepared.
5. Polytron or Dounce homogenizer with a Teflon pestle (0.25 mm clearance, any machine shop can alter the pestle, *see* **Note 3**).
6. Motorized homogenizer.
7. Kim wipes.
8. Scissors (four pairs).
9. 50 ml conical tubes.
10. Buchner funnels fitted with (approximately) 220 and 118 μm meshes with a rubber band and placed onto a flask. The mesh should fit loosely and be slightly dented in the middle forming a trough.
11. 100 mm bacteriological Petri dishes.
12. Glass rods.
13. 45 mm glass beads soaked in HBSS/1 % BSA for at least 1 h (*see* **Note 4**).
14. 60 ml syringes cut at both ends. The sieve is made by securing a piece of nylon mesh (50 μm or greater) around one end of the syringe with a rubber band. The length of the sieve should be approximately 2 cm.

2.2 Permeability Assay

1. Tissue culture medium: Dulbecco's modified Eagle's medium (DMEM) without phenol red enriched with 1× glutamax (Invitrogen, Grand Island, NY, USA), 10 % fetal bovine serum (FBS), and 1× sodium pyruvate.
2. Fluorescein isothiocyanate–dextran (avg wt 70,000) (Sigma Chemical Company, St. Louis, MO, USA).
3. Phosphate-buffered saline (PBS).
4. Mannitol: 1 M in deionized water.
5. Digitonin: 5 % in deionized water (requires heat for solubilization).
6. Oxygen tank.
7. Black 96-well plates.

2.3 *Viability Assay*	1. CytoTox 96 Non-Radioactive Cytotoxicity Assay (Promega, Madison, WI, USA).

2.4 *Hematoxylin and Eosin Staining*	1. Methanol.
	2. Hematoxylin.
	3. Eosin.
	4. Acidified alcohol: 2,578 ml 95 % ethanol, 950 ml deionized water, and 9 ml HCl.

3   Methods

3.1 *Isolating Brain Microvessels*

This method takes advantage of the relative strength of the brain microvessel to mechanical force. The method described here is for isolating brain microvessels from cows, but can be applied to other large mammals. The procedure for rodents is similar, except for the method of disrupting the brain. A polytron is used for larger brains, while a homogenizer is used for smaller ones. Microvessels are separated from other cellular constituents by floating the myelin on a dextran gradient by centrifugation. We have also used dextran for isolating rat brain microvessels, though other labs have used Ficoll [28]. Microvessels adhere to the sieve, whereas nuclei and other cell debris pass through it. An important note is that this procedure does not distinguish capillaries from arterioles and venules. Thus, the procedure is for preparing microvessels not capillaries. Arterioles and venules also express tight junctions and can be considered to comprise the BBB.

3.2 *Dissecting and Homogenizing the Brain*

1. Brains must be obtained fresh from the abattoir and quickly immersed in HBSS that is prechilled in an ice bath. Glassware and centrifuge tubes should also be prechilled. Microvessels are highly prone to loosing ATP. Therefore, all procedures are conducted at 4 °C to maintain high vessel viability.

2. Completely remove the meninges by carefully peeling it from the parenchyma. Remove the brain stem, corpus callosum, and cerebellum. Cut pieces of grey matter into a beaker filled with HBSS and discard the remaining white matter. Mince the brain into 1–2 mm size pieces using four sharp scissors in one hand.

3. Pour the minced brain to the 20 ml mark in a 50 ml conical tube and add 20 ml of HBSS. Brains from larger animals are more conveniently dispersed with a polytron, although a homogenizer can be used. Brains from rodents are dispersed with a homogenizer.

4. Apply a polytron at the lowest setting needed to make a suspension within 20 s. The polytron and homogenizer destroy the cellular integrity of glia and neurons, but the microvessels remain intact. In homogenizing tissue, use a 5:1 ratio of brains to HBSS.

Similar to using the polytron, keep the homogenizer tube on ice to prevent any heat buildup. Also, the number of strokes should be constant throughout the process. Combine the disrupted tissue and centrifuge at $1,000 \times g$ at 4 °C for 20 min.

5. The pellet is not solid, but rather has a consistency similar to a slurry. Measure the volume of the pellet using a graduated cylinder and transfer into a flask. Add dextran to achieve a final concentration of 18 % and mix thoroughly for 1 min. The dextran and homogenates must be well mixed; however, shaking too vigorously will decrease the overall vessel yield. An alternative is not to centrifuge, but to simply combine the dextran with the homogenate. While doing this saves a step, it requires much more dextran for the procedure.

6. Centrifuge to separate myelin from cells and microvessels (*see* **Note 5**). The myelin will be present at the top of the supernatant and the grey matter containing the microvessels will form the pellet. Remove the myelin carefully and pour the supernatant out. There will also be myelin on the walls of the centrifuge tubes, which should be removed with paper towels being careful not to disturb the red pellet. Place the tube on ice and add one volume (volume of the pellet) of ice-cold HBSS/BSA to the pellet. Tap the bottom of the tube vigorously or vortex slowly to resuspend the pellet. Pipeting with a 25 ml pipette is also helpful. The pellet will likely be clumpy. Transfer the suspension to a 50 ml conical tube and polytron for a 10–20 s. Be careful not to contaminate the pellet with myelin.

7. This filtering step removes clumps composed of fat and coagulated blood and larger vessels, which will clog the glass bead sieve. The filter fits loosely on the funnel, so it can be indented and secured with a rubber band. Pipette the suspension slowly onto 210 µm nylon filter covering a Buchner funnel that sits on a flask in an ice bath. Using a glass rod, gently rub the filter to insure a continual flow of fluid. After all of the suspension is added, wash the filter several times with a generous amount of HBSS/BSA.

8. Use the same setup and re-filter the suspension through a 118 µm nylon mesh filter.

9. During the filtering process, make glass bead sieves (*see* **Note 5**). We use two sieves for one bovine brain (about 45 g of tissue). Carefully transfer the filtrate to the sieve. The nuclei and other debris will pass through the sieve, but the microvessels will adhere to the glass beads. After allowing the filtrate to pass through, wash the sieve with 750 ml of HBSS/BSA. Pour the beads into a beaker, pipette enough HBSS/BSA to cover the beads, and swirl to dislodge the microvessels from the beads. After each swirl, allow the beads to settle to the bottom of the beaker and transfer the HBSS/BSA to a fresh 50 ml conical

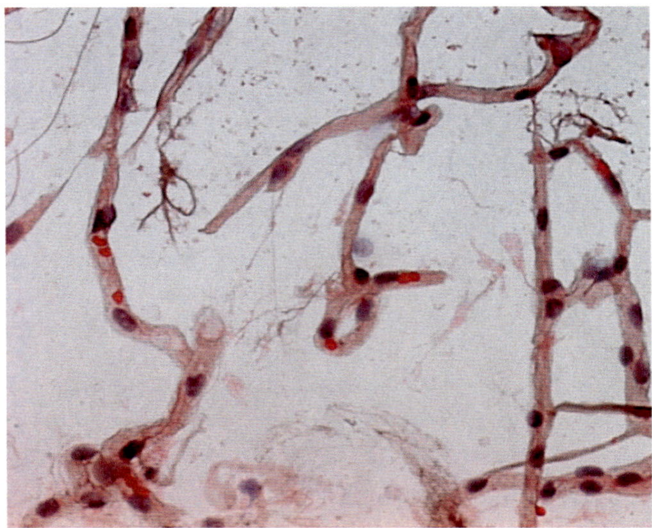

Fig. 1 Brain microvessel preparation stained with hematoxylin and eosin. This batch of BM displayed an 18.5-fold increase in gamma-glutamyl transpeptidase-specific activity compared to controls. The micrograph was taken with a 40× objective

tube. Repeat three times, with the first swirl being gentle and the last one being the most vigorous.

10. Combine the three washings and centrifuge at $1,000 \times g$ for 10 min at 4 °C. Resuspend the pellet in a small volume of HBSS/BSA and place a drop of the suspension onto a microscope slide. Under phase contrast microscopy, microvessels will appear with some debris (*see* **Note 6**). Additional washings may yield greater purity. Let vessels air-dry on slide and apply hematoxylin and eosin to stain (Subheading 3.5).

11. Total yield will generally vary between 18 and 22 mg per 45 g of grey matter.

12. The two criteria for assessing batches of isolated BM are the absence of contaminating nuclei (Fig. 1) and specific activity of gamma-glutamyl transpeptidase greater than 17 U/mg protein. Hematoxylin and eosin staining assesses nuclei through phase contrast microscopy and is used to monitor purity during the isolation. At each step of enrichment, there will be an increase in gamma-glutamyl transpeptidase-specific activity. Alkaline phosphatase activity will also increase, while levels of acetylcholine esterase will decrease. We have found that gamma-glutamyl transpeptidase activity alone is sufficient to assess purity. The procedure has been revised over many years [28, 29] and can also be applied to isolate BM from humans [30].

3.3 Maintaining Microvessel Viability and Cryopreservation

Although BMs display energy-dependent mechanisms and the ability to produce ATP from mitochondria, they readily take up trypan blue. To measure viability, for example in assays for permeability and drug transport, we recommend measuring release of lactate dehydrogenase (LDH). An MTT assay or measuring ATP levels, in our experience, requires more material than the LDH release assay. The percent of LDH release is normalized to total activity. Various protocols have been reported to maintain viability through establishing BM in a perfusion chamber, plating on extracellular matrix, and the use of continual agitation. A constant in each of these protocols is to oxygenate the media, provide a high level of glucose (5 g/l) and 1 mM pyruvate, and maintain the cultures at 37 °C.

1. Wash an aliquot of freshly prepared BMs and resuspend in PBS. Determine total protein content by Bradford assay or other protein assay.

2. Collect BM by centrifugation at $1{,}000 \times g$ for 5 min and resuspend the pellet in culture medium containing 5 % DMSO to a final protein concentration of 2 mg/ml.

3. Freeze 1 ml aliquots slowly overnight at −80 °C. The frozen vessels can be stored at −80 °C.

3.4 Permeability Assay

Tight junctions perform very similar functions in epithelium and the BBB, but the proteins composing tight junctions are different in different cell types. In addition, signaling pathways regulating tight junction integrity are varied in a cell-specific context [31, 32]. Consequently, brain endothelial cells must be examined when studying the BBB and tight integrity, not endothelial cells from another source. Chemicals have been identified that compromise tight junction integrity through interactions with various tight junction proteins [33–35]. In vivo, high concentrations of the sugars, mannitol and arabinose, have been used to temporarily increase the permeability of the BBB [36]. This higher permeability generally lasts for less than 10 min and in BM impermeability returns the mannitol away. The permeability assay developed in our laboratory is described here. The assay's Z′ score of 0.452 approaches excellence in quality assessment and the assay's signal window of 2.8 indicates that the difference between background and treatment with positive control is sufficient to be considered robust [37].

1. Freshly prepare or thaw BM and resuspend at a concentration of 200 μg/250 μl of media without phenol red. Transfer 250 μl of this suspension into one well of a 48-well plate. Plate triplicates for samples and controls.

2. Prepare chemicals of interest and vehicle controls and add to the test wells, bringing the final volume of each well up to 300 μl. Prepare the fluorescein isothiocyanate–dextran (average wt 70,000) in PBS at a concentration of 25 μg/ml and add 10 μl to each well, except those used as background controls.

3. To test suitability of cells for the permeability assay, use three wells and add 0.5 M mannitol. Incubate for 10 min, then wash BM with PBS left on for 10 min, and add FITC–dextran to test both viability and suitability of the BM. The recovery period allows tight junctions to reform and no FITC–dextran should be taken up. Batches of BM are not suitable for the permeability assay if they do not recover from the mannitol treatment and PBS wash.

4. Set up experimental controls: three wells—nothing added to serve as background controls; three wells—add 0.5 M mannitol or 50 μg/ml digitonin; if left on for the duration of FITC incubation this will allow for maximum uptake of FITC–dextran.

5. After a 1-h incubation, wash the BM three times with PBS and centrifuge the cells for 5 min at $1,000 \times g$. Resuspend BM in 100 μl of PBS and snap freeze and thaw prior to transferring into a black 96-well plate. Read fluorescence at an excitation of 494 nm and emission of 521 nm.

3.5 Hematoxylin and Eosin Staining

Hematoxylin and eosin allows for the differential staining of nuclear and cytoplasm, allowing you to examine the gross structure of the vessel. The nuclear material stains blue/purple, while the proteins in the vessel will stain pink.

1. Drop 50 μl of BM suspension onto a glass coverslip and allow to air-dry.

2. Fix the vessels by dipping the slide into ice-cold methanol for 5 min.

3. Dip slide in hematoxylin solution. The length of exposure to hematoxylin will depend on the degree of color desired. Usually we dip the slides for 5 min.

4. Rinse in water and then dip once into acidified alcohol.

5. Rinse in water and dip into eosin solution for 15–30 s.

6. Allow the slide to dry and mount with a coverslip.

7. Visualize the microvessels by light microscopy.

3.6 Cell Culture Model and Measuring Transendothelial Electrical Resistance

The advantage of a cell culture model is that transport can be measured in both the abluminal-to-luminal and luminal-to-abluminal directions and that their gene expression potential can be altered. Models using primary cultures of bovine [38], porcine [4], and rat brain endothelial cells [39] display mean electrical resistances approaching $1,000 \, \Omega/cm^2$. Endothelial cell lines may also be used, though none display electrical resistances approaching $1,000 \, \Omega/cm^2$. The rat RBE4 [1] and human HCMEC/D3 cell lines [40] have been used in many different laboratories. Almost all cell culture models are validated by measuring TER, which evaluates tight junction integrity [41].

Whichever culture model the investigator chooses, it is imperative to validate the integrity of the tight barrier. A starting point is measuring TER because tight junctions will not permit paracellular transport of ions. TER measurements need to be conducted at the beginning of an assay to verify that the cells are fully differentiated, and at the termination of the experiment to verify that the tight junctions remain intact. Several devices, both commercial and manufactured in-house, have been reported to measure TER. Here, chopstick electrodes will be discussed because they are less expensive than other devices and widely used. Measurements are made in the medium used to grow cells or in a balanced salt solution, such as HBSS.

1. Pre-warm the HBSS or medium in advance of the procedure.

2. Add the same volume of HBSS at 37 °C into the wells of the plate (same size as those used for filter inserts).

3. Prepare one extra well than the number of inserts to be tested.

4. To transfer the inserts into the new well, gently decant the media on the apical side by removing the insert from the plate with forceps. Add a fixed amount of the pre-warmed HBSS to the apical side of each insert.

5. To determine the blank, the level of resistance is measured in an insert without cells. This insert must be treated identically to that with cells. For example, inserts with cells might have been coated with extracellular matrix; therefore the blank insert must also be coated with the same extracellular matrix.

6. The cells and blank are equilibrated before measuring resistance by placing the plate in the incubator for 20 min.

7. During the equilibration time, place the chopstick electrodes in 70 % ethanol for 15 min to sterilize.

8. Remove the electrodes from ethanol in a sterile environment, let dry briefly, and then place them into a 15 ml centrifuge tube filled with enough pre-warmed HBSS to bathe the electrode tips.

9. Set the meter to resistance (or ohms) and connect to the electrodes.

10. Test the meter according to meter instructions, usually by pressing the "Test" button or similar.

11. Measure the resistance by placing chopstick electrodes into the plate so that the longer chopstick rests on the bottom of the well plate and the shorter one rests on the apical section of the insert.

12. Press the measurement button on the meter and record the resistance in each well. We suggest taking three measurements per well to ensure accuracy.

13. Because the measurements are taken under aseptic conditions, the media can be added back to the cells and the plate can be placed in the incubator for any additional studies.

14. The level of resistance is calculated by subtracting the average resistance in the blank insert from the average resistance of the insert with cells. This is multiplied by the surface area (in cm^2) of the insert and reported as ohms-cm^2. Additionally, changes in resistance following a treatment can be calculated by comparing changes in experimental wells (before and after exposure) to changes in untreated wells over the same period of time.

3.7 Paracellular Transport and Permeability

1. Culture cells and plate on transwell filters as described above.

2. Measure experimental variables in duplicate or triplicate.

3. Wash chambers three times with phenol red-free HBSS, taking care not to disturb the monolayer.

4. Add HBSS to the basolateral chamber to the level of the membrane and to cover the apical chamber. Place the plate in the incubator at 37 °C for 10 min. Initiate transport by adding sodium fluorescein to achieve a final concentration of 10 µg/ml.

5. Remove the HBSS from the chamber opposite from the one receiving dyes into separate containers at desired time points. Be sure to store the samples in foil to protect them from quenching by light exposure and at –20 °C if they are not going to be analyzed immediately.

6. Analyze the samples using a spectrophotometer at 490 nm or spectrofluorometer at an excitation wavelength in the range of 440–480 nm and an emission wavelength of 517 nm. Transport is measured by the equations stated above in Subheading 3.6.

4   Notes

1. The amount of BSA can vary between 0.5 and 1.0 %. In the absence of BSA, brain microvessels will bind to the tube and other surfaces.

2. To make HBSS with a final dextran concentration of 30 %, slowly dissolve dextran (sizes 60,000–120,000 kDa) in water to make a 35 % solution. Add 1 ml of 10× HBSS to every 8.5 ml of 35 % dextran solution and 0.5 ml of HEPES (1 M, pH 7.4). It is imperative that the dextran solution is made correctly.

3. A machine shop can shave the pestle to a 0.25 mm clearance. Each homogenizing tube and pestle will be matched and used only for isolating BM.

4. The size of the glass beads is 25 mm in most procedures, but we have found that 25 mm glass beads clog more frequently than 45 mm beads. The flow should be relatively fast, so adjust accordingly. Many procedures describe this step as glass bead columns, rather than sieves, which is the term used in this chapter. We suggest that the term column is a misnomer and does not describe the true objective of using glass beads. The longer microvessels will wrap around the beads and remain attached, while cell debris and nuclei will have a greater tendency to flow through. We suggest that the 25 and 45 mm beads be compared for vessel yield and purity.

5. To separate myelin from cell bodies, centrifugation speeds vary from 5,000 to 25,000×g. Many investigators report better yields at the higher speeds. We have also found better yields at the higher centrifugation speed, but we also have encountered problems with resuspending the pellet. An alternative is to collect the pellet at the lower speed, collect the supernatant, resupsend the myelin again, and recentrifuge at the higher speed. Both pellets can then be recombined. We also recommend a brief (several seconds) polytron step to disperse any clumps.

6. Tight junctions on microvessels close very fast after mannitol is withdrawn and fluorescent compounds will remain trapped in the microvessels after washings. Interestingly, BM incubated with 0.5 M mannitol and dextran conjugated to fluorescein together and washed retained fluorescence. Thus, the opening of the tight junctions is transient. The procedure could potentially be used to trap macromolecules in the lumen.

References

1. Roux F, Couraud PO (2005) Rat brain endothelial cell lines for the study of blood–brain barrier permeability and transport functions. Cell Mol Neurobiol 25:41–58

2. Gumbleton M, Audus KL (2001) Progress and limitations in the use of in vitro cell cultures to serve as a permeability screen for the blood–brain barrier. J Pharm Sci 90: 1681–1698

3. Aschner M, Fitsanakis VA, dos Santos AP et al (2006) Blood–brain barrier and cell-cell interactions: methods for establishing in vitro models of the blood–brain barrier and transport measurements. Methods Mol Biol (Clifton, NJ) 341:1–15

4. Patabendige A, Skinner RA, Abbott NJ (2013) Establishment of a simplified in vitro porcine blood-brain barrier model with high transendothelial electrical resistance. Brain Res 1521:1–15

5. Milton SG, Knutson VP (1990) Comparison of the function of the tight junctions of endothelial cells and epithelial cells in regulating the movement of electrolytes and macromolecules across the cell monolayer. J Cell Physiol 144:498–504

6. Handler JS (1983) Use of cultured epithelia to study transport and its regulation. J Exp Biol 106:55–69

7. Crone C, Olesen SP (1982) Electrical resistance of brain microvascular endothelium. Brain Res 241:49–55

8. Lippmann ES, Azarin SM, Kay JE et al (2012) Derivation of blood–brain barrier endothelial cells from human pluripotent stem cells. Nat Biotechnol 30:783–791

9. Cucullo L, Couraud PO, Weksler B et al (2008) Immortalized human brain endothelial cells and flow-based vascular modeling: a marriage of convenience for rational neurovascular studies. J Cereb Blood Flow Metab 28:312–328

10. Felix RA, Barrand MA (2002) P-glycoprotein expression in rat brain endothelial cells: evidence for regulation by transient oxidative stress. J Neurochem 80:64–72

11. Seetharaman S, Barrand MA, Maskell L et al (1998) Multidrug resistance-related transport proteins in isolated human brain microvessels and in cells cultured from these isolates. J Neurochem 70:1151–1159

12. Perriere N, Yousif S, Cazaubon S et al (2007) A functional in vitro model of rat blood–brain barrier for molecular analysis of efflux transporters. Brain Res 1150:1–13

13. Bachmeier CJ, Trickler WJ, Miller DW (2006) Comparison of drug efflux transport kinetics in various blood–brain barrier models. Drug Metab Dispos 34:998–1003

14. Shapiro AB, Ling V (1998) The mechanism of ATP-dependent multidrug transport by P-glycoprotein. Acta Physiol Scand 643:227–234

15. Polgar O, Robey RW, Bates SE (2008) ABCG2: structure, function and role in drug response. Expert Opin Drug Metab Toxicol 4:1–15

16. Wang Q, Strab R, Kardos P et al (2008) Application and limitation of inhibitors in drug-transporter interactions studies. Int J Pharm 356:12–18

17. Yang Z, Horn M, Wang J et al (2004) Development and characterization of a recombinant madin-darby canine kidney cell line that expresses rat multidrug resistance-associated protein 1 (rMRP1). AAPS J 6:77–85

18. Wolff JE, Belloni-Olivi L, Bressler JP et al (1992) Gamma-glutamyl transpeptidase activity in brain microvessels exhibits regional heterogeneity. J Neurochem 58:909–915

19. Vernon H, Clark K, Bressler JP (2011) In vitro models to study the blood–brain barrier. Methods Mol Biol (Clifton, NJ) 758:153–168

20. Shivers RR, Betz AL, Goldstein GW (1984) Isolated rat brain capillaries possess intact, structurally complex, interendothelial tight junctions; freeze-fracture verification of tight junction integrity. Brain Res 324:313–322

21. White FP, Dutton GR, Norenberg MD (1981) Microvessels isolated from rat brain: localization of astrocyte processes by immunohistochemical techniques. J Neurochem 36:328–332

22. Erdlenbruch B, Alipour M, Fricker G et al (2003) Alkylglycerol opening of the blood–brain barrier to small and large fluorescence markers in normal and C6 glioma-bearing rats and isolated rat brain capillaries. Br J Pharmacol 140:1201–1210

23. Hartz AM, Bauer B, Fricker G et al (2004) Rapid regulation of P-glycoprotein at the blood–brain barrier by endothelin-1. Mol Pharmacol 66:387–394

24. Bauer B, Hartz AM, Lucking JR et al (2008) Coordinated nuclear receptor regulation of the efflux transporter, Mrp2, and the phase-II metabolizing enzyme, GSTpi, at the blood–brain barrier. J Cereb Blood Flow Metab 28:1222–1234

25. Hartz AM, Mahringer A, Miller DS et al (2010) 17-beta-Estradiol: a powerful modulator of blood–brain barrier BCRP activity. J Cereb Blood Flow Metab 30:1742–1755

26. Zhang Y, Schuetz JD, Elmquist WF et al (2004) Plasma membrane localization of multidrug resistance-associated protein homologs in brain capillary endothelial cells. J Pharmacol Exp Ther 311:449–455

27. Miller DS, Nobmann SN, Gutmann H et al (2000) Xenobiotic transport across isolated brain microvessels studied by confocal microscopy. Mol Pharmacol 58:1357–1367

28. Dallaire L, Tremblay L, Beliveau R (1991) Purification and characterization of metabolically active capillaries of the blood–brain barrier. Biochem J 276(Pt 3):745–752

29. Goldstein GW, Wolinsky JS, Csejtey J et al (1975) Isolation of metabolically active capillaries from rat brain. J Neurochem 25:715–717

30. Hargreaves KM, Pardridge WM (1988) Neutral amino acid transport at the human blood–brain barrier. J Biol Chem 263:19392–19397

31. Forster C (2008) Tight junctions and the modulation of barrier function in disease. Histochem Cell Biol 130:55–70

32. Anderson JM, Van Itallie CM (1995) Tight junctions and the molecular basis for regulation of paracellular permeability. Am J Physiol 269:G467–G475

33. Haorah J, Heilman D, Knipe B et al (2005) Ethanol-induced activation of myosin light chain kinase leads to dysfunction of tight junctions and blood–brain barrier compromise. Alcohol Clin Exp Res 29:999–1009

34. Yeh TH, Hsu LW, Tseng MT et al (2011) Mechanism and consequence of chitosan-mediated reversible epithelial tight junction opening. Biomaterials 32:6164–6173

35. McCall IC, Betanzos A, Weber DA et al (2009) Effects of phenol on barrier function of a human intestinal epithelial cell line correlate

with altered tight junction protein localization. Toxicol Appl Pharmacol 241:61–70

36. Rapoport SI (2000) Osmotic opening of the blood–brain barrier: principles, mechanism, and therapeutic applications. Cell Mol Neurobiol 20:217–230

37. Iversen PW, Eastwood BJ, Sittampalam GS et al (2006) A comparison of assay performance measures in screening assays: signal window, Z′ factor, and assay variability ratio. J Biomol Screen 11:247–252

38. Parran DK, Magnin G, Li W et al (2005) Chlorpyrifos alters functional integrity and structure of an in vitro BBB model: co-cultures of bovine endothelial cells and neonatal rat astrocytes. Neurotoxicology 26:77–88

39. Abbott NJ, Dolman DE, Drndarski S et al (2012) An improved in vitro blood–brain barrier model: rat brain endothelial cells co-cultured with astrocytes. Methods Mol Biol (Clifton, NJ) 814:415–430

40. Weksler BB, Subileau EA, Perriere N et al (2005) Blood–brain barrier-specific properties of a human adult brain endothelial cell line. FASEB J 19:1872–1874

41. Griepp EB, Dolan WJ, Robbins ES et al (1983) Participation of plasma membrane proteins in the formation of tight junctions by cultured epithelial cells. J Cell Biol 96: 693–702

Chapter 7

Methods to Assess Tissue Permeability

Juan C. Ibla and Joseph Khoury

Abstract

An essential requirement for adequate organ performance is the formation of permeability barriers that separate and maintain compartments of distinctive structure and function. The endothelial cell lining of the vasculature defines a semipermeable barrier between the blood and interstitial spaces of all organs. Disruption of the endothelial cell barrier can result in increased permeability and vascular leak. These effects are associated with multiple systemic disease processes and can accompany acute tissue responses to injury. The mechanisms that control barrier function are complex and their full understanding requires a multidisciplinary approach. The use of in vivo permeability data often complements molecular findings and adds power to the studies. The interaction of multiple cell types and tissues present only in mammalian models allows for testing of hypothesis and establishing the physiological significance of the results. In this chapter we describe simple methods that can be used systematically to measure the permeability profile of several organs.

Key words Permeability assay, Vasculature, Edema, Inflammation, Evans blue dye, Fluorescence bioparticles, Water content, Wet-to-dry ratio

1 Introduction

Mucosal surfaces provide a physical barrier between the environment and the internal layers of many organs. The dual role of maintaining close contact yet separated biological compartments requires multiple interactions between endothelia, epithelia, and transmembrane adhesion domains. This highly adapted cellular hierarchy denotes a specific anatomical location within an organ (e.g., polarized epithelia) and assigns individual cell types vital tasks for organ function. The endothelial lining of all organs is charged to interact with the systemic circulation, transduce soluble messages, and control regional blood flow and organ development [1, 2]. Insight into the mechanisms that regulate these natural barriers is necessary for the understanding of normal physiology and disease progression. It is well established that abnormal barrier function is present in a multitude of disease states and that available methods can be used to estimate the severity of diseases and their response to treatment [3].

Troy A. Baudino (ed.), *Cell-Cell Interactions: Methods and Protocols*, Methods in Molecular Biology, vol. 1066,
DOI 10.1007/978-1-62703-604-7_7, © Springer Science+Business Media New York 2013

Several methods have been developed to study endothelial permeability in the vasculature, both in vivo and in vitro [4]. These include blood clearance and diffusion methods that can be performed both invasively and noninvasively in animals and human subjects [5]. In this book chapter, we describe an easy, clearly interpretable, and highly reproducible method to assess tissue permeability that can be readily accomplished with equipment available in standard cell biology laboratory settings.

2 Materials

2.1 Miles Assay for Vascular Permeability

1. Mice.
2. Phosphate-buffered saline (PBS).
3. Evans blue dye.
4. Specific ligand, cytokine, growth factor, or condition being tested.
5. 1 ml syringe with needles.
6. Surgical dissecting equipment (e.g., scissors, forceps).
7. Photographic equipment.
8. Formamide and 10 % buffered formalin.
9. Water bath set at 55 °C.
10. Spectrophotometer at 450 nm and 590–620 nm.
11. 96-well plate or cuvette for spectrophotometer.

2.2 Fluorescent-Labeled Molecules

1. Mice.
2. PBS.
3. Fluorescent isothiocyanate (FITC)–dextran, molecular weight 4–40 kDa (Sigma-Aldrich, St. Louis, MO, USA), or fluorescent-labelled microspheres (Molecular Probes, Eugene, OR, USA).
4. 1 ml syringes and 23-G needles.
5. 1.5 ml microcentrifuge tubes.
6. Fluorescent plate reader.
7. Fluorescent microscope.
8. Plates for fluorescence.

2.3 Labeled Bioparticles

1. Fluorescent bacteria commercially labeled with fluorescein (Alexa Fluor, Life Technologies, Grand Island, NY, USA).
2. Fluor (488/594) and Texas Red dye among others (Molecular Probes).
3. Sterile PBS.

4. 2 nM sodium azide.

5. Vortex mixer.

6. Hemocytometer.

7. Fluorescence microscope.

8. Mice.

2.4 Water Content in Tissues

1. Microcentrifuge tubes.

2. High-fidelity electronic balance.

3. Speed-Vac centrifuge.

4. Tissue of interest.

3 Methods

3.1 Miles Assay for Vascular Permeability

The classical method in which vascular permeability can be measured in vivo is by the Miles assay for vascular permeability, also known as the Evans blue dye method [6]. Evans blue is a small-molecule marker that strongly associates with albumin, allowing for quantification of vascular leakage into the extravascular tissues. The accumulated dye can be quantified by the use of a spectrophotometer. Studies using this technique have been used to understand the molecular mechanisms involved in abnormal barrier function during inflammation [7], ischemia–reperfusion injury [8], and also as a quantitative end point to test therapeutic interventions that modulate barrier function [9]. Results are often evident at the macroscopic level, as depicted in Fig. 1.

1. Reconstitute Evans blue to a final concentration of 0.5 % (w/v) in PBS.

2. Administer 0.2 ml of Evans blue by tail vein injection to each mouse.

3. The treatment or the condition of choice is then administered to the mice, with a subset of mice being left untreated to serve as controls.

4. Humanely sacrifice the mice as per institutional guidelines.

5. Dissect out the organs/tissues of interest and photograph as necessary.

6. Quantify the Evans blue concentration by eluting 50 mg of the individual tissue in 0.5 ml of formamide at 55 °C for 2 h to overnight.

7. Measure aliquots of the eluted dye by absorbance at 610 nm with subtraction of the reference absorbance at 450 nm using a spectrophotometer.

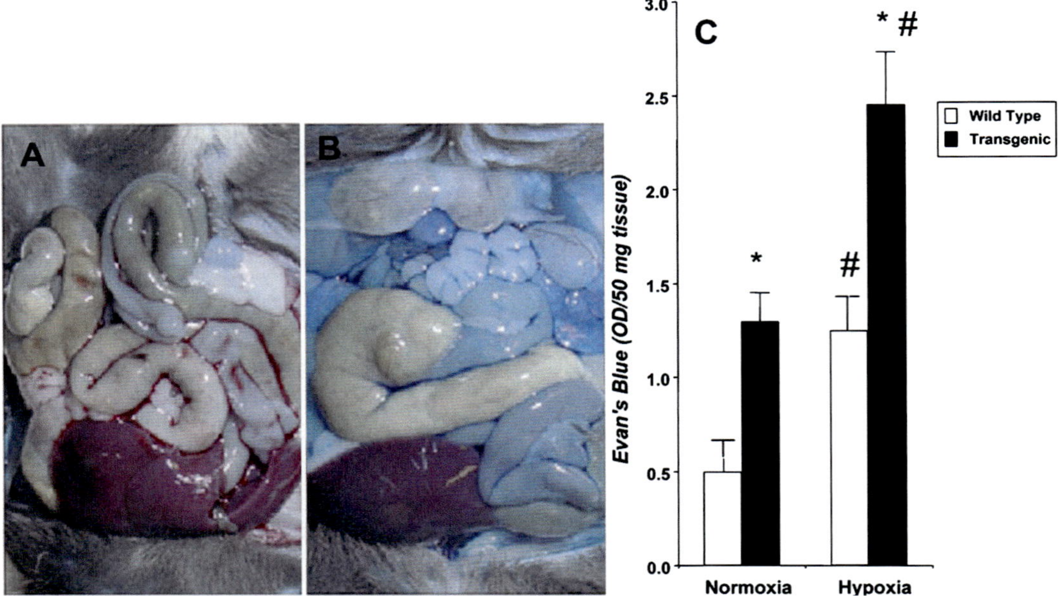

Fig. 1 Photographs representing normal (**a**) and increased (**b**) vascular permeability. C57Bl6 male mice were injected with 0.2 ml of Evans blue dye via tail vein. Animals were then subjected to 4 h of normobaric hypoxia (8 % O_2, 92 % N_2) or room air conditions. Postmortem macroscopic examination of intra-abdominal viscera displayed significant differences between experimental and control groups. (**c**) Evans blue concentration (OD/50 mg of tissue) in lung tissue of normal and transgenic mice with increased pulmonary vascular permeability. *$p < 0.05$ compared to wild-type mice, #$p < 0.01$ compared to normal oxygen conditions

3.2 Fluorescent-Labeled Molecules

Fluorescent dyes are linked to either dextran or polysterene microspheres that can be used as tracers for permeability. These tracers may then be visualized either directly in vivo by the aid of intravital microscopy or in tissue sections. Several dyes and molecule sizes may be used to assay both epithelial and endothelial permeability. Although the initial method employed by Rudolph and Heymann [10] used radiolabeled microspheres, growing concerns over environmental and health issues as well as cost of handling, disposal, and half-life of the radiolabel have brought about the development of non-isotope fluorescent or colored molecules. Fluorescent-labeled molecules are routinely employed for permeability assays using both endothelial and epithelial cells [11] and more recently to understand real-time tissue responses to immunologic stimuli [12]. In addition, fluorescent-labeled particles can be used to detect molecules in solutions and solid samples [13]. The method we describe here uses FITC–dextran [14], but it may easily be adapted for various fluorescent-labeled microspheres (*see* **Note 1**).

1. Reconstitute the FITC–dextran in PBS to a concentration of 80 mg/ml.

2. Administer the FITC–dextran to each animal at a dose of 60 mg/100 g of body weight (*see* **Note 2**).

3. Expose animals to the condition of interest for the appropriate amount of time (2–6 h).

4. Collect whole blood by cardiac puncture in anesthetized animals using the syringe and needle.

5. Transfer blood to a 1.5 ml microcentrifuge tube, let sit on ice for 1 h, spin at $7,000 \times g$ for 5 min, and collect the supernatant (serum).

6. Using a fluorescence plate, make serial dilutions of FITC–dextran in PBS (range from 10 to 0.015 μg/ml), and to unknown wells, add 60 μl of 1:33 diluted serum in PBS per well (triplicate).

7. Read the plate at excitation wavelength 485 nm and emission wavelength 530 nm.

8. Measurements are recorded as nanograms of FITC–dextran per microliter of serum.

3.3 Labeled Bioparticles

The dynamic interaction of cellular components within a tissue can be analyzed in vivo by the kinetics of labeled bioparticles during exposure to critical physiological conditions. This technique utilizes a series of fluorescent-labeled, heat or chemically killed bacteria of a variety of sizes and antigenic characteristics. These methods have been employed to study a diverse group of parameters including epithelial permeability, phagocytosis, and opsonization. Results can be obtained by fluorescent microscopy, quantitative spectrophotometry, and flow cytometry [15, 16]. We have used these methods to study bacterial translocation in transgenic mice exposed to mucosal barrier-disruptive conditions (hypoxia).

1. Reconstitute labeled bacteria (100 mg, 3×10^8 *Escherichia coli*) to 20 mg/ml in sterile PBS (*see* **Note 3**).

2. Add 2 mM sodium azide to the bacterial suspension and store at 4 °C. Without the addition of sodium azide, the bacterial suspension should be used within 1 day.

3. Vigorously vortex the solution three times for 15 s at the highest setting.

4. Count the number of bioparticles per ml using a hemocytometer and a phase contrast or a fluorescent microscope.

5. Dilute the bioparticles to 3×10^7 per 0.5 ml.

6. Vigorously vortex the solution three times for 15 s at the highest setting.

7. Using a 1 ml syringe, inject the bioparticle suspension into mice via oral/esophageal injection.

8. Expose the animals to the desired experimental conditions.

9. At various time points (1–24 h), harvest target organs/tissues and serum.

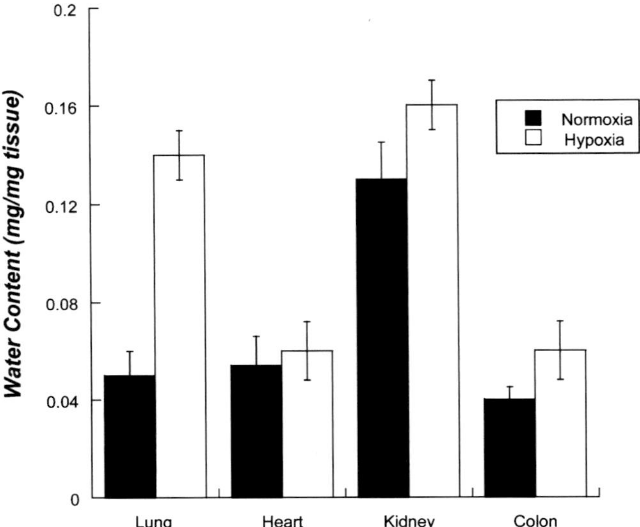

Fig. 2 Vascular permeability profile of different organs in mice. All organs were harvested and water content quantified by wet/dry ratios. Multiple organs were profiled simultaneously from experimental animals exposed to normal oxygen and hypoxic conditions. Data are expressed in mg/mg of tissue

10. Fluorescence can be quantified directly in serum or by eluting organ aliquots in formamide at 55 °C from 2 h to overnight.

11. Dissected organs can be fixed in formalin, embedded in paraffin, and cut for fluorescence microscopy.

3.4 Water Content in Tissues

Tissue inflammation is often characterized by cellular infiltration and accumulation of extracellular water (edema). Quantifying water content in tissues can be a useful measurement to determine the integrity of vascular and epithelial barriers. This parameter can often be combined with other permeability assays to corroborate the effect of experimental conditions or pharmacological interventions [17, 18]. We describe a simple and highly accurate method to measure water content in tissue using dry-to-wet weight ratios. Multiple organs can be assayed simultaneously as shown in Fig. 2.

1. After the experimental conditions have been completed, collect 20–50 mg of tissue of interest from a representative portion of the organ (*see* **Note 4**).

2. Obtain the wet weight of the organ using an electronic high fidelity balance and record (tissue alone and tube containing tissue).

3. Place the samples in labeled microcentrifuge tubes (make sure that the sample is placed at the bottom of the tube).

4. Place the tubes in a speed-vac and set up for vacuum mode at 65 °C for 12–18 h (*see* **Note 5**).

5. Obtain the dry weight of the tissue.

6. Subtract the dry weight from wet weight to obtain total water content.

4 Notes

1. If using fluorescent microspheres, fluorescence may be measured directly in serum or in frozen sections of tissues using a fluorescence microscope.

2. For intestinal epithelial permeability, FITC–dextran solution is given by oral gavage. For pulmonary epithelial permeability, FITC–dextran is injected directly into the trachea of anesthetized animals.

3. Bioparticles are provided as lyophilized powders. Upon receipt, they should be stored frozen at −20 °C. In the case of fluorescent-labeled products, these should be protected from light exposure.

4. Smaller amounts of tissue (10–30 mg) from organs with high fat content (intestine) or high cellular density (kidney) can be used, since these tissues will take longer to dry.

5. If tissues from different organs are processed simultaneously, check the vacuum run periodically (every 6–8 h). Lung tissue, depending on the amount, will typically require less time to dry compared to other organs. Overexposure of the samples may result in protein evaporation, tissue destruction, and inaccurate readings.

References

1. Beck KF, Eberhardt W, Frank S et al (1999) Inducible NO synthase: role in cellular signalling. J Exp Biol 202:645–653

2. Bertuglia S, Giusti A (2005) Role of nitric oxide in capillary perfusion and oxygen delivery regulation during systemic hypoxia. Am J Physiol Heart Circ Physiol 288: H525–H531

3. Hofmann T, Stutts MJ, Ziersch A et al (1998) Effects of topically delivered benzamil and amiloride on nasal potential difference in cystic fibrosis. Am J Respir Crit Care Med 157:1844–1849

4. Karhausen J, Ibla JC, Colgan SP (2003) Implications of hypoxia on mucosal barrier function. Cell Mol Biol (Noisy-le-grand) 49:77–87

5. Bleeker-Rovers CP, Boerman OC, Rennen HJ et al (2004) Radiolabeled compounds in diagnosis of infectious and inflammatory disease. Curr Pharmal Design 10:2935–2950

6. Miles AA, Miles EM (1952) Vascular reactions to histamine, histamine-liberator and leukotaxine in the skin of guinea-pigs. J Physiol 118:228–257

7. Cui B, Sun JH, Xiang FF et al (2012) Aquaporin 4 knockdown exacerbates streptozotocin-induced diabetic retinopathy through aggravating inflammatory response. Exp Eye Res 98:37–43

8. Liu K, Sun T, Wang P et al (2013) Effects of erythropoietin on blood–brain barrier tight junctions in ischemia-reperfusion rats. J Mol Neurosci 49:369–379

9. Chintagari NR, Liu L (2012) GABA receptor ameliorates ventilator-induced lung injury in rats by improving alveolar fluid clearance. Crit Care 16:R55

10. Rudolph AM, Heymann MA (1967) The circulation of the fetus in utero. Methods for studying distribution of blood flow, cardiac output and organ blood flow. Circ Res 21: 163–184

11. Rosenberger P, Khoury J, Kong T et al (2007) Identification of vasodilator-stimulated phosphoprotein (VASP) as an HIF-regulated tissue permeability factor during hypoxia. FASEB J 21:2613–2621

12. Lu M, Munford RS (2011) The transport and inactivation kinetics of bacterial lipopolysaccharide influence its immunological potency in vivo. J Immunol 187:3314–3320

13. Wei SC, Hsu PH, Lee YF et al (2012) Selective detection of iodide and cyanide anions using gold-nanoparticle-based fluorescent probes. ACS Appl Mater Interfaces 4:2652–2658

14. Synnestvedt K, Furuta GT, Comerford KM et al (2002) Ecto-5′-nucleotidase (CD73) regulation by hypoxia-inducible factor-1 mediates permeability changes in intestinal epithelia. J Clin Invest 110:993–1002

15. Samel S, Keese M, Kleczka M et al (2002) Microscopy of bacterial translocation during small bowel obstruction and ischemia in vivo—a new animal model. BMC Surg 2:6

16. Sorrells DL, Friend C, Koltuksuz U et al (1996) Inhibition of nitric oxide with aminoguanidine reduces bacterial translocation after endotoxin challenge in vivo. Arch Surg 131:1155–1163

17. Thompson LF, Eltzschig HK, Ibla JC et al (2004) Crucial role for ecto-5′-nucleotidase (CD73) in vascular leakage during hypoxia. J Exp Med 200:1395–1405

18. Toung TJ, Chang Y, Lin J et al (2005) Increases in lung and brain water following experimental stroke: effect of mannitol and hypertonic saline. Crit Care Med 33:203–208, discussion 259–260

Chapter 8

In Vivo Quantification of Metastatic Tumor Cell Adhesion in the Pulmonary Microvasculature

F. Bartsch, M.L. Kang, S.T. Mees, J. Haier, and P. Gassmann

Abstract

In vivo and ex vivo fluorescence video microscopy used to be a well-established method in life science with a variety of applications, such as in inflammation or cancer research. In this book chapter, we describe a model of in vivo fluorescence microscopy of the rat's lung with the exclusive advantage of qualitative and quantitative in vivo analysis of cell adhesion within the complex microenvironment of the ventilated and perfused lung. Observation can include real-time, time-lapse, or fast-motion analysis. In our laboratory, we have used the model for qualitative and quantitative real-time analyses of metastatic colon cancer cell adhesion within the rat's pulmonary microcirculation. Using some modifications in another series, we have also applied the model to analyze thrombocyte and leucocyte adhesion within the pulmonary capillaries in experimental sepsis. For interventional studies, injected cells or animals may be pretreated with various reagents or drugs for further analysis of adhesion molecules involved in tumor cell–endothelial cell interactions.

Key words In vivo fluorescence video microscopy, Cancer metastases, Tumor cell adhesion, Microvasculature, Endothelial cells, Lung, Rat

1 Introduction

Side-specific adhesive interactions of circulating cells are well established for leucocytes in the capillaries of inflamed tissue and a large number of surface molecules as well as intracellular signalling cascades have been described controlling these cell–cell and cell–matrix interactions [1]. Recently, organ-specific adhesive interactions of circulating tumor cells within the capillaries of metastatic target organs have been recognized to contribute to the organ-specific nature of cancer metastasis formation [2]. In analogy to inflammation biology, the phenomenon of side-specific metastasis formation was called "homing" of metastasizing tumor cells to their specific metastatic target organ.

For example, one-third of patients diagnosed with colorectal cancer will eventually die from this disease representing approximately 600,000 deaths worldwide per year [3]. Most of these

Troy A. Baudino (ed.), *Cell-Cell Interactions: Methods and Protocols*, Methods in Molecular Biology, vol. 1066, DOI 10.1007/978-1-62703-604-7_8, © Springer Science+Business Media New York 2013

89

colorectal cancer-related deaths are due to distant metastatic growth, rather than local tumor progression. As in almost any cancer, the pattern of metastatic growth is nonrandom and in the case of colorectal cancer, liver and lung are the predominant metastatic targets beside lymph nodes and the peritoneal cavity [4]. The organ-specific characteristic of metastatic growth has already been recognized by S. Paget [5] in the late nineteenth century and his "seed-and-soil concept" has been modified by additional findings and dissected into single steps which are outlined in the literature as the "metastatic cascade" [6]. Based on various in vivo and ex vivo techniques, several groups suggested mechanically restricted tumor cell embolism in the first capillary bed entered by circulating tumor cells [6, 7]. In contrast, our group [8] and others [9] have demonstrated organ-specific tumor cell adhesion in the microcirculation of the liver and lung [10–12].

Experimental metastasis models are widely used in cancer research. Cells are usually injected via the intravenous or intraarterial route. In end point assays, after sacrifice of the animal, the numbers of metastases of single arrested cells are quantified within a specific organ by conventional histology or immunohistochemistry. In interventional studies, the number of induced metastases can then be modified by certain interventions, for example by inhibition of adhesion molecules on the cells' surface. Nevertheless, the precise mechanisms of such interventions remain unclear in endpoint metastasis assays. On the other hand, in vitro models like static or flow adhesion assays are useful to evaluate the adhesive actions of certain molecules and their modulation by intracellular signalling cascades [13]. However, in vitro it is obvious that tumor cell adhesion is a complex process, influenced by not only the expression of adhesion molecules but also other factors such as mechanical sheer stress [14]. Furthermore, it has been demonstrated that the local chemokine profile can promote tumor cell adhesion and migration and contribute to organ-specific metastasis formation by chemotaxis [15].

In our experimental studies, human colon cancer cells showed an organ-specific pattern of cell arrest in rats mimicking the clinical picture of metastatic colon cancer [16] and Glinksii et al. [12] reported organ-selective metastatic tumor cell arrest of human breast cancer cells in a mouse model following the typical clinical pattern of metastatic breast cancer diseases. Circulating tumor cells adhere to microvascular endothelial cells (EC) and subendothelial extracellular matrix (ECM) proteins by different sets of adhesion molecules. Subsequent and regulated trans-endothelial cell migration taking place at secondary sites in response to several microenvironmental factors has also been reported [8–11, 15, 17].

In an attempt to quantitatively and qualitatively analyze metastatic tumor cell arrest and to further dissect the mechanisms of metastatic tumor cell adhesion in pulmonary capillaries, we developed this in vivo model using in vivo fluorescence microscopy of the rat lung. Furthermore, this model has also proved valuable for the analyses of leucocyte and thrombocyte adhesion in experimental sepsis [18].

2 Materials

2.1 Cell Culture and Cell Preparation

For maintenance of tissue cell culture and preparation of single-cell suspensions for injection, use your standard laboratory protocols and equipment. For certain steps provide the following:

1. Ca^{2+}–Mg^{2+}-free phosphate-buffered solution (CMF-PBS).
2. Bovine serum albumin (BSA).
3. 15 and 50 ml conical tubes.
4. Cell counter or hemacytometer.
5. CalceinAM.
6. Fluorescine (FITC)-labeled dextran 10,000 kDa.

2.2 Instruments and Catheters

1. Forceps.
2. Scissors.
3. Microsurgical forceps.
4. Microsurgical scissors.
5. Needle holder.
6. Bipolar forceps.
7. Syringes (1 ml).
8. Indwelling venous catheters (14 and 22 G) (*see* **Note 1**).
9. Microsurgical vascular clamps.
10. Polyfilament sutures 4-0.

2.3 Small Animal Surgical Room Equipment

For the operative procedure, the following equipment for a small animal surgical room is recommended (*see* **Note 1**):

1. Dissecting microscope.
2. Isoflurane/N_2O vaporizer and supply.
3. Small animal respirator.
4. Inhalation chamber for anesthesia induction.
5. Oro-nasal overtube for inhalation anesthesia (e.g., 20 ml syringe).
6. Heating-plate or bio-feedback platform for temperature maintenance of the animal.
7. Device for bipolar coagulation.
8. Upright in vivo fluorescence microscope including 20× magnification.

9. Corresponding fluorescence filters to the fluorescence dyes used (in our studies, the most commonly used is fluorescein).

10. For in vivo microscopy, prepare an individually designed holder for a thin glass coverslip (*see* Fig. 4). The in vivo microscope should be equipped with a camera and an analog or a digital recording system. Though not required, a blood gas analysis system is also useful for these studies.

2.4 Animals

Housing and maintenance of the animals as well as the experimental procedures must be in accordance with the institute and country's regulations. The protocol must be approved by the local authorities and/or animal welfare committee, if necessary. In our laboratory, male Sprague Dawley rats (CD rats) or athymic, T cell-deficient RNU rats (Charles River) were used, but other strains of rats or mice specific to individual labs may be used as well (*see* **Note 2**).

2.5 Cells

Most of our studies involve the use of human colon cancer cell lines (*see* **Note 2**); however, different cell types relevant to individual labs can be used.

1. The highly metastatic subclone HT-29LMM (I. Fidler, Houston, Texas, USA) was derived from repeated in vivo passaging of parental HT29 cells [19].

2. T84 cells (ATCC, Manassas, VA, USA) are a transplantable human carcinoma cell line derived from a lung metastasis of a human colon carcinoma.

3. For different purposes, we also used labeled leucocytes or platelets for adhesion experiments.

3 Methods

3.1 Workflow Overview

One should start preparing the surgical room and the instruments needed for the procedure. After this, we recommend preparation of the cells that will be used (as a single-cell suspension). During the time of cell reconstitution (approximately 45 min), the surgical procedure can be performed. After completion of the operation and placement of the animal under the microscope, the cells are washed and adjusted to the desired cell concentration. If you work in a team of two or more persons, you may adjust this workflow accordingly (Fig. 1). The total time needed, including preparation of the surgical suite, is about 4 h for one experiment.

3.2 Cell Preparation

1. Depending on the aim of the study, you may rinse the cells with CMF-PBS for up to 24 h prior to the experiment.

2. Culture the cells with medium containing 1 % BSA instead of FBS, to exclude the influence of growth factors from the FBS on adhesion molecule expression on the cell surface.

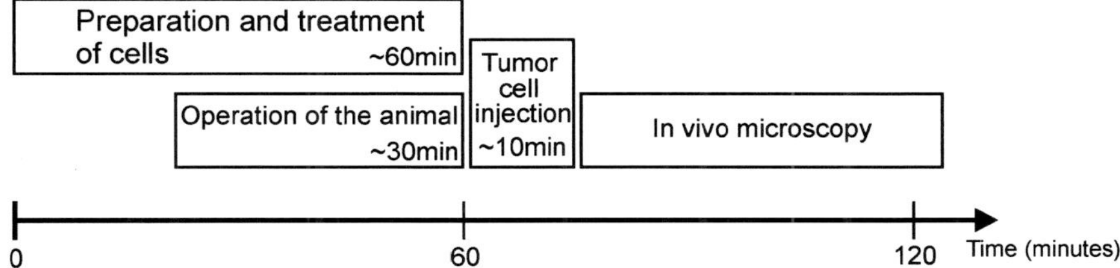

Fig. 1 Experimental procedure. The experiment starts with the trypsinization, reconstitution, and treatment of the cells. This takes approximately 60 min. While cells are incubating, proceed with the operation of the animal. The surgery should take about 30 min. After incubation of the cells, the animal is positioned under the microscope, cells are injected, and in vivo microscopy is performed. *Bottom*: Timeline in minutes

3.3 Preparing the Single-Cell Suspension

1. Cells are trypsinized by standard culture protocols and washed once in CMF-PBS.

2. Resuspend cells in FBS-free medium containing 1 % BSA.

3. Add CalceinAM (1 µg/ml) and keep cells as a single-cell suspension for 45 min in the incubator (37 °C, 5 % CO_2) for reconstitution of cell surface molecules.

4. Mix the cells gently over the incubation period using an orbital shaker or by hand every few minutes.

5. For blockade of adhesion molecules or other interventions, incubate cells during the reconstitution period with inhibiting or activating antibodies or other substances to be studied.

6. After 45 min, wash cells in CMF-PBS, resuspend in CMF-PBS, and adjust to a concentration of 1×10^6 cells/ml.

7. Inject the cells as a single-cell suspension in a volume of approximately 1 ml over a 1–2-min period of time.

8. Proceed to in vivo microscopy (Subheading 3.5).

3.4 Animal Preparation

1. *Anesthesia and surgical procedure*: Place animals in a chamber with a volume of approximately 1 l and induce anesthesia by insufflation of isoflurane 5 %/N_2O with an oxygen fraction (FiO_2) of 0.33 %.

2. After induction of anesthesia, weigh animals and fix in a supine position on a heating platform set at 37 °C to maintain body temperature. At this point, maintain anesthesia by inhalation of isoflurane 2.5 %/N_2O via an overtube placed on the animal's mouth and nose.

3. *Placement of a permanent arterial catheter* (*see* **Note 3**): Shave and disinfect the throat and thorax of the animal.

4. Remove a skin-flap from the anterior aspect of the throat.

5. For the placement of a central-arterial catheter, use a dissecting microscope and mobilize the right submandibular gland and isolate the right carotid artery from beneath the sterno-cleidomastoid muscle.

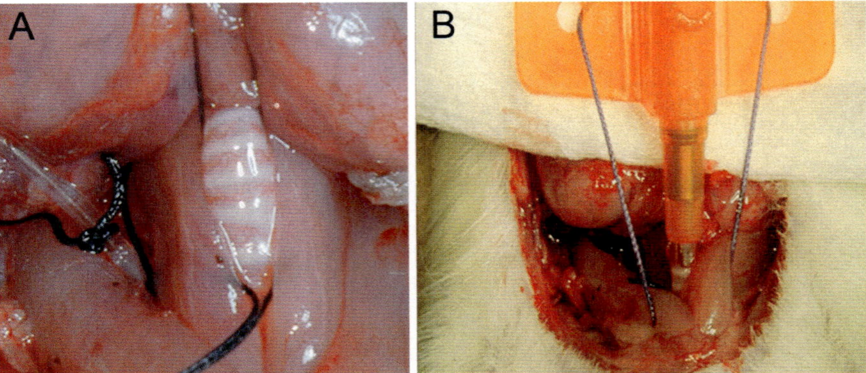

Fig. 2 Animal preparation #1. Two general steps of the animal preparation are presented. (**a**) A catheter is inserted into the right arteria carotis and fixed with sutures. In the center, the trachea is exposed and a suture is prepared for later fixation of the tracheal-cannula. (**b**) After incision of the trachea and insertion of the ventilation-cannula, the cannula is fixed with a suture to the origin of the musculi sternocleidomastoidei

6. Place one suture proximal and one suture distally around the artery.

7. After placement of a clamp centrally, the distal part is ligated.

8. Using microsurgical scissors, perform an arteriotomy and introduce a catheter adjusted to the artery's diameter (approximately 22 G) into the artery and place with the tip centrally to the heart.

9. Fix the catheter by ligation of the prepared sutures (Fig. 2a).

10. Gently flush the catheter with saline.

11. *Establishment of mechanical ventilation*: Using the same skin incision, isolate the trachea by separation of the ventral muscles and place a suture just beneath it (Fig. 2a).

12. After performing a tracheotomy, use a 14 G indwelling catheter as a tracheal cannula and fix it with an additional suture to the origin of the sternocleidomastoid muscle at the sternum (Fig. 2b). The length of the catheter may by adjusted before placement.

13. After tracheotomy and placement of the tracheal cannula, the connection to the respirator must be performed in a rapid sequence to ensure continuous anesthesia and ventilation to the animal.

14. Mechanical ventilation should be performed at a rate of 35–40/min and a tidal volume of 3.5–4.0 ml for CD rats weighing approximately 250 g. These parameters, especially the tidal volume, need to be adjusted to the specific animal's body weight. Adequate ventilation can be confirmed by arterial blood gas analysis at the end of the experiment.

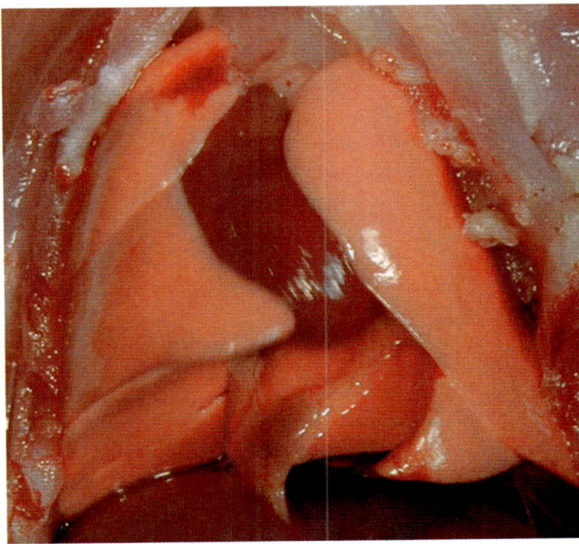

Fig. 3 Animal preparation #2. Picture of the chest of the animal after thoracotomy. The lungs are exposed and prepared for microscopy without any atelectasis

15. *Thoracotomy:* For in vivo microscopy, remove the anterior aspect of the chest using scissors and bipolar coagulation (Fig. 3).

16. It is mandatory to avoid any contact to the lungs. Touching the lung's surface immediately causes atelectasis and precludes in vivo microscopy of this area. For well-trained researchers, it is possible to remove only the right lower part of the thoracic wall, as for the microscopy procedure only the right lower lobe of the lung is used.

3.5 In Vivo Microscopy

1. Place the prepared animal under an upright in vivo fluorescence microscope (*see* **Note 4**).

2. For fluorescence in vivo microscopy, mount a thin glass coverslip on a specially designed holder and carefully place upon the lung's surface with minimal pressure to avoid development of any atelectasis (Fig. 4; *see* **Note 5**).

3. Position the animal in a left-lateral position to minimize moving artifacts during ventilation (*see* **Note 6**).

4. We use a 20× magnification objective with water immersion. You may use objectives with other specifications according to the aim of your study.

5. The lung's surface must be irrigated permanently with 0.9 % saline.

6. For positive contrast of the blood vessels, slowly inject the animal with 500 µg FITC-labeled dextran through the permanent catheter. Using a similar protocol, earlier studies have

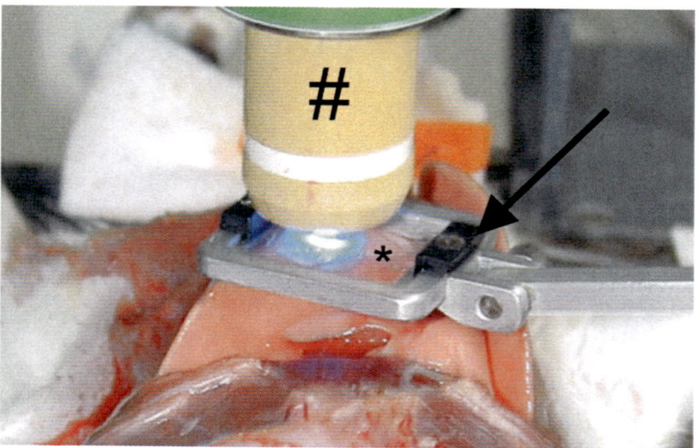

Fig. 4 In vivo microscopy of the lung. The animal is positioned on the left lateral site for better exposition of the right lung. A thin glass coverslip (*asterisk*) mounted on a specially designed holder (*arrow*) is carefully placed upon the lung's surface with minimal pressure. Through the glass cover, in vivo microscopy is performed with a 20× magnification objective (*hash*) using water immersion

demonstrated stable hemodynamic conditions and constant microcirculation in the lungs for at least 60 min [18].

7. Inject 1×10^6 CalceinAM-labeled cells as a single-cell suspension in 1 ml CMF-PBS over 1 min. Gently flush the catheter (*see* **Note 7**).

8. Standardized lung surface microscopy begins immediately after the injection has been completed.

9. Continuously screen the lung surface in a meander-like fashion recording the number of single adherent tumor cells and the number of microscopic fields (in inspiration) for each time interval analyzed. One area is observed for 10 s and tumor cells adhering for ≥10 s are defined as "stably adherent" (Fig. 5). Emboli of cell clumps are excluded from analysis (*see* **Note 8**).

10. Perform microscopy in 10-min intervals for a total of 40 min so that the observed area is screened four times.

11. Additional findings like passing cells or detaching cells may be recorded as well.

12. At the end of the observation, draw period arterial blood from the left ventricle for a blood gas analysis and sacrifice the animal by potassium injection.

13. The results can be presented as "adherent cells/20 microscopic fields" for each 10-min time interval and can be expressed as means ± standard deviation from "n" independent experiments (animals injected) (*see* **Note 9**).

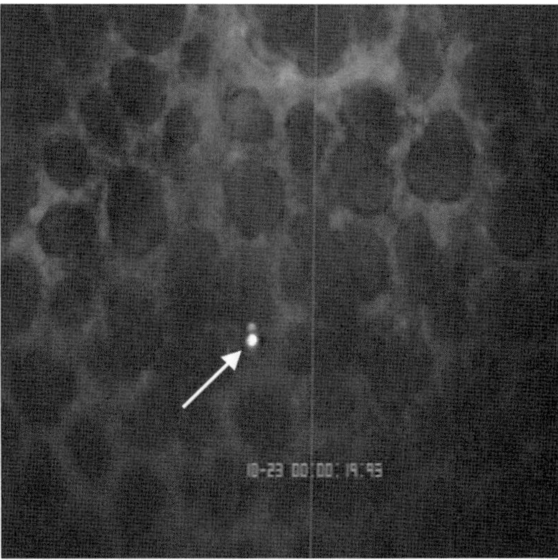

Fig. 5 Picture of in vivo microscopy. Example of an adherent colon cancer cell (*arrow*) in a pulmonary capillary observed by in vivo microscopy (CalceinAM labeled)

To the best of our knowledge, this is one of very few in vivo microscopy models of the rodent lung that enables real-time in vivo observation of cell–capillary interaction. Nevertheless, by this method we are only able to observe subpleural, superficial capillaries on the lung's surface. By in vivo microscopy, we can usually penetrate the tissue for 300 μm but deeper capillaries cannot be observed. It is also not possible to estimate the absolute number of cells arrested in the lungs. The method is therefore semiquantitative and the results always need to be compared to a control group. PET-based techniques may be more helpful at this point [20]. On the other hand, the model presented here allows real-time observation of single cells within the vital pulmonary vasculature. By this technique we have found qualitative differences in the adhesion mechanism between liver and lung. While in liver sinusoids we never found rolling or sticking of colon cancer cells, we observed this behavior in the lung, pointing to different adhesion mechanisms in the two organs [21]. By further analysis of involved adhesion molecules we could exclude tumor cell–ECM interactions in the early onset of colon cancer cell adhesion in the lung, a mechanism that is obviously very early taking place in the liver. There are several papers in the literature that propose tumor cell arrest in the metastatic target organs by size restriction of cells within the capillaries [6, 7]. By in vivo microscopy of the liver using CD rats (250 g) and RNU rats (125 g) we demonstrated that the size of liver sinusoids does not impact the number of arrested cells [22]. In the study mentioned above, using precisely the same technique of lung microscopy as presented here, we were able to interfere with tumor cell–endothelial cell interactions by inhibition of

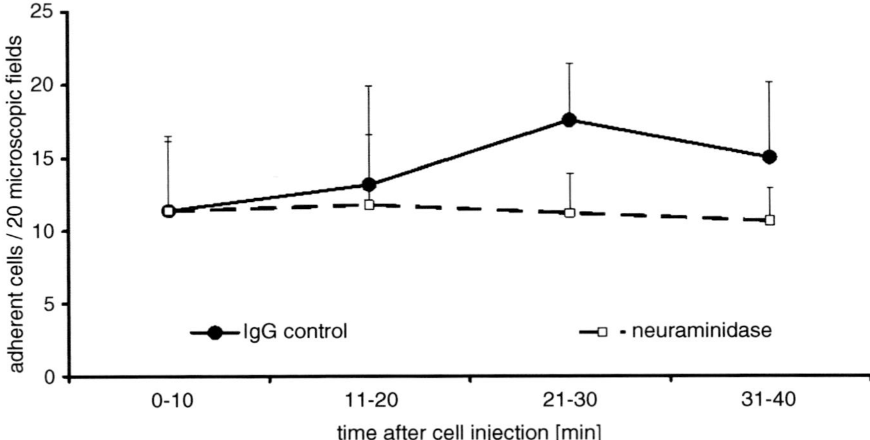

Fig. 6 Inhibition of tumor cell adhesion by glycoprotein digestion. Example of an inhibition experiment. Untreated T cell-deficient RNU rats were injected with 10^6 CalceinAM-labeled cells via the right carotid artery. In the intervention group, ($n=6$) cells were pretreated with neuraminidase for enzymatic digestion of sialylated glycoproteins during the reconstitution period. After 30–40 min, within the pulmonary capillaries significantly less tumor cells were adherent per 20 microscopic fields compared to control cells treated with nonspecific IgG ($n=6$) [30 min: 17.5 ± 4.0 vs. 10.2 ± 2.1, $p=0.005$; 40 min: 15.0 ± 3.2 vs. 10.6 ± 2.9, $p=0.028$]

selectin-mediated cell adhesion supporting the hypothesis of specific adhesive interactions in the early phase of metastatic organ colonization [20]. Figure 6 gives another example of reduced tumor cell adhesion in the pulmonary microvasculature after inhibition of E-selectin-mediated adhesion by enzymatic digestion of sialylated glycoproteins on the tumor cells.

Our group also reported on early tumor cell extravasation of adherent tumor cells from the liver sinusoids. Using in vivo microscopy we were not able to detect cell extravasation in the lung. We are not sure if this is a real finding or whether this is a limit of the technical aspects of local image resolution. There is a considerable debate on the intravascular origin of lung metastases. The observation time of our model is limited to a maximum of 40–60 min. Therefore it allows us only to analyze the very early events of organ colonization. Using this model we are not able to make any statement of later events like eventual extravasation or intravascular proliferation as proposed by Al-Mehdi et al. [10].

4 Notes

1. The described protocol is the way we perform the procedure using our equipment. There are different ways to perform anesthesia, animal maintenance, and catheterization. If your laboratory has different equipment or other standard operating procedures they can work as well and may be maintained for these studies.

2. Other animal strains or cell lines may be used following this protocol.

3. To investigate tumor cell adhesion in the lungs it may be plausible to inject the tumor cells via the venous route as it is used in many experimental metastasis models. In earlier work, we tested the role of the application route in the case of in vivo microscopy of the liver. We found no differences in the cells' behavior, but found the results more reproducible and stable when injecting the cells via the arterial route. Furthermore, injection of the cells via the portal vein caused major hemodynamic disturbances [22]. When transferring the method to the lungs, we experienced the same phenomenon with better reproducible results and more stable hemodynamic conditions of the animals by arterial injection.

4. There may be concerns about the role of size mismatch of tumor cells and the pulmonary capillaries. In earlier studies we determined the median pulmonary diameter in different animal strains. In CD rats (250 g) we found a mean diameter of the pulmonary capillaries of 14.3 μm ± 2.5 μm compared to 8.6 μm ± 2.0 μm in RNU rates weighting 120 g. Comparing the number of adhering cells in the liver sinusoids between the two rat strains we could not detect any significant impact of the different capillary diameters in the liver [23]. Besides this, the humoral immune system does not obviously interfere with tumor cell arrest in this early phase of tumor cell arrest. After injection of weight-adjusted number of cells, we found a similar number of adhering cells in immunocompetent and T cell-deficient rats [23].

5. One major problem in performing in vivo microscopy of the lung is the hemodynamic stability of the animals. It is basically possible to establish full hemodynamic monitoring by invasive blood pressure measurement via the arterial catheter, as well as central venous pressure monitoring by an additional central venous catheter. We demonstrated hemodynamic stability by using the protocol described herein [18]. Nevertheless, for establishment of this technique it may be useful to expand hemodynamic monitoring until the procedure is performed on a regular basis within the laboratory. In our hands, the animals are hemodynamically stable for at least 60 min.

6. One major handicap of the presented model is moving artifacts of the ventilated lung. Therefore, positioning the animal in the left lateral position to minimize the artifacts is crucial, although movement is inevitable. We recommend standardized analysis of each microscopic field in inspiration.

7. The glass coverslip we used in the tumor cell experiments as artificial pleura for immersion microscopy was replaced by a

ring exercising a slight negative pressure on the lung surface in our sepsis studies [18]. This technique delivers less moving artifacts, but offers a smaller surface area for analysis. Nevertheless, it is crucial to apply minimal pressure to the lung because any disturbance during the preparation or the measurement results in atelectasis and this area is lost for analysis.

8. In our laboratory we injected a relatively low number of cells compared to other authors using much higher cell numbers for injection and often smaller volumes. The preparation of a real single-cell suspension may be impossible and cell aggregates may be unavoidable. However, we feel that using fewer cells in a larger volume (1 ml) together with iterative resuspension during reconstitution gives reproducible results.

9. In our experience it takes about 10 min to screen the lung's surface once. One may repeat this step four times. We therefore recommend the presentation of the results in four 10-min intervals. As you will not analyze exactly the same number of microscopic fields in each 10-min interval, the number of adherent cells for each interval needs to be standardized, for example to "cells/20 microscopic fields." The cell numbers given are the number of adherent cells per 20 microscopic fields for each 10-min time interval and are expressed as mean ± standard deviation from "n" independent experiments (animals injected).

References

1. Rose DM, Alon R, Ginsberg MH (2007) Integrin modulation and signaling in leukocyte adhesion and migration. Immunol Rev 218:126–134

2. Gassmann P, Haier J (2008) The tumor cell—host interface in the early onset of metastatic organ colonisation. Clin Exp Metastasis 25:171–181

3. Shibuya K, Mthers CD, Boschi-Pinto C et al (2002) Global and regional estimates of cancer mortality and incidence by site: II. Results for the global burden of disease 2000. BMC Cancer 2:37

4. Weiss L, Grundmann E, Torhorst J et al (1986) Haematogenous metastatic patterns in colonic carcinoma: an analysis of 1541 necropsies. J Pathol 150:195–203

5. Paget S (1889) The distribution of secondary growth in cancer of the breast. Lancet 1:571–573

6. Chambers AF, Groom AC, MacDonald IC (2002) Dissemination and growth of cancer cells in metastatic sites. Nat Rev Cancer 2:563–572

7. Mook OR, Van Marle J, Vreeling-Sindelarova H et al (2003) Visualization of early events in tumor formation of eGFP-transfected rat colon cancer cells in liver. Hepatology 38:295–304

8. Enns A, Korb T, Schluter K et al (2005) Alphavbeta5-integrins mediate early steps of metastasis formation. Eur J Cancer 41:1065–1072

9. Brodt P, Fallavollita L, Bresaller RS et al (1997) Liver endothelial E-selectin mediates carcinoma cell adhesion and promotes liver metastasis. Int J Cancer 71:612–619

10. Al-Mehdi AB, Tozawa K, Fisher AB et al (2000) Intravascular origin of metastasis from the proliferation of endothelium-attached tumor cells: a new model for metastasis. Nature 6:100–102

11. Zipin A, Israeli-Amit M, Meshel T et al (2004) Tumor-microenvironment interactions: the fucose-generating FX enzyme controls adhesive properties of colorectal cancer cells. Cancer Res 64:6571–6578, Erratum in: Cancer Res 2004, 64:8130

12. Glinskii OV, Huxley VH, Glinsky GV et al (2005) Mechanical Entrapment is insufficient and intracellular adhesion is essential for metastatic cell arrest in distant organs. Neoplasia 7:522–527

13. Haier J, Nasralla M, Nicolson GL (2000) Cell surface molecules and their prognostic values in assessing colorectal carcinomas. Ann Surg 231:11–24

14. Haier J, Nicolson GL (2000) Tumor cell adhesion of human colon carcinoma cells with different metastatic properties to extracellular matrix under dynamic conditions of laminar flow. J Cancer Res Clin Oncol 126:699–706

15. Müller A, Homey B, Soto H et al (2001) Involvement of chemokine receptors in breast cancer metastasis. Nature 410:50–56

16. Schluter K, Gassmann P, Enns A et al (2006) Organ-specific metastatic tumor cell adhesion and extravasation of colon carcinoma cells with different metastatic potential. Am J Pathol 169:1064–1073

17. Voura EB, Ramjeesingh RA, Montgomery AMP et al (2001) Involvement of integrin $\alpha v \beta 3$ and cell adhesion molecule L1 in tranendothelial migration of melanoma cells. Mol Biol Cell 12:2699–2710

18. Küpper S, Mees ST, Gassmann P et al (2007) Hydroxyethyl starch normalizes platelet and leukocyte adhesion within pulmonary microcirculation during LPS-induced endotoxemia. Shock 28:300–308

19. Price JE, Daniels LM, Campbell DE et al (1989) Organ distribution of experimental metastases of a human colorectal carcinoma injected in nude mice. Clin Exp Metastasis 7:55–68

20. Kikkawa H, Kaihou M, Horaguchi N et al (2002) Role of integrin alpha(v)beta3 in the early phase of liver metastasis: PET and IVM analyses. Clin Exp Metastasis 19:717–725

21. Gassmann P, Kang ML, Mees ST et al (2010) In vivo tumor cell adhesion in the pulmonary microvasculature is exclusively mediated by tumor cell-endothelial cell interaction. BMC Cancer 10:177

22. Haier J, Korb T, Hotz B et al (2003) An intravital model to monitor steps of metastatic tumor cell adhesion within the hepatic microcirculation. J Gastrointest Surg 7:507–514, discussion 514–5

23. Gassmann P, Hemping-Bovenkerk A, Mees ST et al (2009) Metastatic tumor cell arrest in the liver-lumen occlusion and specific adhesion are not exclusive. Int J Colorectal Dis 24:851–858

Chapter 9

Cell Membrane Vesicles as a Tool for the Study of Direct Epithelial–Stromal Interaction: Lessons from CD147

Eric Gabison, Farah Khayati, Samia Mourah, and Suzanne Menashi

Abstract

Communication between the epithelial and stromal tissue layers, separated by basement membrane, is known to provide the information necessary for development, differentiation, and homeostasis. These interactions are altered in benign or malignant diseases, in particular when the basement membrane barrier is disrupted allowing a greater proximity between the two cell layers that triggers tissue remodeling. Epithelial–stromal interactions (ESI) have been examined in vitro by various approaches that can be broadly divided into interactions arising from secreted diffusible factors and interactions through direct cell–cell contact. Here we describe a method for the study of direct ESI through CD147, an adhesion molecule present on the epithelial cell surface and which is known to interact with stromal cells, such as fibroblasts and endothelial cells, and signal them to increase production of matrix metalloproteinases. This method can be extended to other adhesion molecules involved in ESI.

Key words Tumor–stroma interactions, CD147, Membrane vesicles, Stromal cells, Epithelial cells

1 Introduction

Basement membrane separates epithelia from the underlying connective tissue and is thought to participate in mediating and modulating epithelial–stromal interactions (ESI). Its disruption modifies the nature of these interactions in a way that triggers stromal remodeling. During wound healing, this step represents the initiation of a repair process that aims to restore tissue homeostasis. In cancer, basement membrane rupture reflects the point where localized tumor cells progress to a more invasive state which also permits direct interactions between the tumor cells and the stroma.

ESI can be mediated through soluble factors, such as cytokines and growth factors, which are small diffusible factors that permit communication between cells located at a distance. Most studies in this field have focused on the biological effects of cytokine produced by one cellular compartment (stroma or epithelium) on the cells from the other compartment. In this chapter we focus on the

Troy A. Baudino (ed.), *Cell-Cell Interactions: Methods and Protocols*, Methods in Molecular Biology, vol. 1066,
DOI 10.1007/978-1-62703-604-7_9, © Springer Science+Business Media New York 2013

103

study of direct contact between epithelial cells and stromal cells (direct-ESI), which may occur once the basement membrane separating the two compartments is disrupted. Such interactions require close proximity between the cells and involve recognition between cell surface proteins.

The notion of direct-ESI was first introduced in the cancer field when CD147 (also referred to as EMMPRIN) was identified as an inducer of matrix metalloproteinases (MMPs), present on the cell surface of tumor cells which can activate stromal cells through direct contact and signal them to increase MMPs' production [1–3]. Immunohistochemical analysis detected CD147 mainly at the periphery of invasive tumor clusters, corresponding to the leading edge of tumor invasion and compatible with the concept that CD147 plays a role in tumor–stroma interactions [4]. Since then, CD147 has been shown to modulate MMP expression during non-tumoral pathological situations as well as in normal tissue remodeling and differentiation [2].

Although ubiquitously expressed, CD147 appears to be enriched on the surface of epithelial cells. We have previously shown that CD147 expressed in normal corneal epithelium is able to activate corneal fibroblasts in culture through direct contact with epithelial cells and to trigger MMP production and myofibroblast differentiation, suggesting a role for CD147 as a key mediator of direct-ESI during corneal wound healing [5–7].

CD147, also commonly referred to as EMMPRIN or basigin, is a membrane glycoprotein that belongs to the immunoglobulin (Ig) superfamily [1]. It is composed of two C2-like immunoglobulin domains, a transmembrane domain, and a short cytoplasmic domain [8]. Using tagged expression vectors and cross-linking experiments, CD147 molecules have been shown to associate with each other on the plasma membrane, forming homo-oligomers in a *cis*-dependent manner, with the potential to increase the overall avidity of CD147 on the cell surface and its ability to induce MMP production in neighboring cells through direct interaction [9, 10].

In order to study the effect of epithelial cell surface CD147 on the host stromal cells, we analyze fibroblasts after incubation with CD147-enriched membrane vesicles. With this system, CD147 is integrated within membrane vesicles and hence represents best the natural spatial configuration in which it exerts its activity. The alternative option of using the purified protein has additional disadvantages as the purification of native CD147, being an insoluble transmembrane protein, is difficult and the recombinant CD147 that is commercially available only contains the soluble extracellular domain lacking the transmembrane domain. This method using membrane vesicles also presents an advantage over epithelial/stromal cell coculture systems which contain both cell types and therefore exert bidirectional effects, while with using CD147-enriched membranes, any synthetic activity or alteration in cellular physiology can be attributed to the effect of CD147 on fibroblasts alone. However, using CD147-containing membranes derived from epithelial cells has the drawback that they constitutively express high levels of CD147 on their surface membrane so that a negative control of membranes in which CD147 is absent is difficult to obtain. For that reason, we use Chinese hamster ovary (CHO) cells which allow, by cDNA transfection, us to obtain membrane vesicles that contain CD147 or lack CD147 expression. CHO cells were chosen since they can be efficiently transfected using standard protocols and have consistently provided successful expression for many of our proteins of interest.

These CHO-derived CD147-containing membranes, but not the control CD147-deficient membranes, increased the production of several MMPs in fibroblasts and in other cell types such as endothelial cells and tumor cells [11, 12] confirming that CD147 is functionally active within these membrane vesicles. Here we describe the method of membrane preparation from CD147 and mock-transfected CHO cells and their incubation with fibroblasts.

2 Materials

2.1 Cell Culture

1. Tissue culture flasks 75 cm^2.

2. Culture medium for CHO cells: Dulbecco modified Eagle's medium (DMEM)/F12 supplemented with 2 ml glutamine and penicillin/streptomycin (100 U/ml).

3. Culture medium for fibroblasts: DMEM supplemented with 10 % fetal bovine serum (FBS), 2 ml glutamine, and penicillin/streptomycin (100 U/ml).

4. Fetal calf serum (FCS).

5. 0.05 % Trypsin–EDTA.

6. Sterile phosphate-buffered saline (PBS) pH 7.4.

2.2 Membrane Preparation

1. Tissue culture flasks 75 cm^2.

2. DMEM/F12 serum free.

3. Protease inhibitor cocktail (Roche, Madison, WI, USA), containing a mixture of several protease inhibitors for the inhibition of serine and cysteine, but not metalloproteases.

4. Sterile 15 ml conical tubes.

5. Sterile PBS pH 7.4.

6. Cell scrapers.

7. Centrifuge tubes, thick-wall polycarbonate, 1 ml, 11 × 34 mm.

2.3 Incubation of Fibroblasts with Membrane Vesicles from CHO–CD147 and CHO–Mock-Transfected Cells

1. Sterile PBS pH 7.4.
2. Lysis buffer NP40: NaCl (150 mM), NP-40 (1 %), and Tris–HCl (50 mM, pH 8.0). For 1 l of NP-40 lysis buffer, combine 30 ml of 5 M NaCl, 100 ml of 10 % NP-40, 50 ml of 1 M Tris (pH 8.0), and 820 ml of deionized water. Store at 4 °C.
3. Trizol (Life Technologies, Grand Island, NY, USA).
4. 6-well plates.
5. 1.5 ml centrifuge tubes.

2.4 Gelatin Zymography

1. Gelatin stock solution: 10 mg/ml gelatin in double-distilled water. Warm up the water in a water bath at 37 °C to dissolve the gelatin, aliquot, and store at –20 °C. Before use, thaw the aliquot in a 37 °C water bath.
2. Zymogram gel: 10 % SDS-polyacrylamide gel with 1 mg/ml gelatin incorporated (*see* **Note 1**).
3. 4× Laemmli sample buffer: 270 mM Tris–HCl pH 6.8, 8 % SDS, 40 % glycerol, 0.004 % bromophenol blue (do not add β-mercaptoethanol or DTT which inactivates MMPs).
4. 5× Running buffer: 125 mM Tris and 960 mM glycine. Store at room temperature.
5. 2.5 % (v/v) Triton X-100.
6. Zymogram developing buffer: 50 mM Tris–HCl, pH 7.6, 5 mM $CaCl_2$, 0.02 % NaN_3, and 0.1 % Triton X-100.
7. Coomassie staining buffer: 0.2 % Coomassie Brilliant Blue, 10 % glacial acetic acid, and 30 % methanol.
8. Destaining solution: 10 % glacial acetic acid and 10 % methanol.

3   Methods

3.1 Cell Culture

CHO cells are stably transfected with CD147 cDNA in the pCR3.1 plasmid or mock transfected using empty plasmid. These cells are then used for the preparation of membrane vesicles. These CHO-derived membrane vesicles are incubated with the fibroblasts of interest. In our study, we generally use HTK hTert-immortalized corneal fibroblasts (a kind gift from J. Jester, University of Texas, Austin, TX, USA) [13]. The hTert-immortalized foreskin fibroblasts can be used as skin dermal cells. Alternatively, primary fibroblasts can be used after isolation from skin explants.

1. Culture CHO cells in DMEM/F12 with 10 % FCS until acquiring an adequate number of cells for experiments.
2. Culture fibroblasts in DMEM with 10 % FCS until acquiring enough cells for experiments.

3. Cells (CHO and fibroblasts) are cultured at 37 °C in humidified atmosphere of 5 % CO_2 and passaged using trypsin–EDTA when approaching confluence (80–90 %). Cells are generally split 1:3.

3.2 Membrane Preparation from the Transfected CHO Cells

1. Plate the CHO–CD147 and CHO–mock-transfected (empty vector) cells in T75 flasks until approximately 80 % confluence (*see* **Note 2**).

2. Aspirate the medium, wash plates 2× in sterile PBS to remove serum and dead cells, and aspirate PBS (*see* **Note 3**).

3. Add 2 ml of serum-free DMEM/F12 medium supplemented with protease inhibitors (4 ml of inhibitor cocktail for 100 ml of medium) to each T75 flask.

4. Place the flask on ice in a sterile cell culture hood.

5. Scrape the cells using a scraper previously cleaned with water and ethanol or use a freshly opened sterile cell scraper. Make sure to scrape all areas and corners of the flask for maximum yield. The medium will appear cloudy.

6. Transfer this suspension to a sterile 15 ml conical tube. Combine suspensions when using several flasks at a time. Keep suspension on ice at all times.

7. Sonicate the suspension using three 10-s cycles at 40 W on ice. The sonicator probe should be washed with 70 % ethanol to maximize sterility. It is recommended to wear ear protection while sonicating.

8. Centrifuge the cell lysate at $1,000 \times g$ for 10 min at 4 °C (to remove unbroken cells and large cellular debris).

9. Transfer the supernatant to the 1 ml Beckman centrifuge tubes and cover them with parafilm. Tubes and parafilm cuts can be sterilized prior to use by placing them under UV light for 10 min in a sterile hood. The rotor cover should be cleaned with 70 % ethanol. Centrifuge for 30 min at $19,000 \times g$ at 4 °C. This will sediment the granules such as mitochondria and lysosomes and leave the membrane vesicles in the supernatant.

10. Transfer the membrane-containing supernatant to clean and sterilize 1 ml Beckman tube, cover with parafilm, and centrifuge for 1 h at $100,000 \times g$ at 4 °C. Tubes should be completely filled to avoid their collapse at the high g force. This step sediments the membrane vesicles into a transparent pellet (*see* **Note 4**).

11. After centrifugation, carefully remove tubes and place them on ice in a sterile hood. Carefully remove the clear supernatant and resuspend the membrane pellet in 50 μl of cold PBS. The pellet appears as small colorless circle which is highly adherent to the bottom of the tube and requires scratching with a sterile pipette cone and numerous up and down pipetting to resuspend.

The suspension should look milky (depending on the membrane quantity), but translucent.

12. Measure protein concentration by Bradford assay.

13. Use the freshly prepared membranes as described below (do not freeze!).

3.3 Incubation of Fibroblasts with Membrane Vesicles from CHO–CD147 and CHO–Mock-Transfected Cells

1. Culture fibroblasts in 6-well plates or in individual 35 mm dishes.

2. When approximately 80 % confluence is achieved, wash twice with serum-free DMEM to remove serum.

3. Add 20 μg/ml of the membrane suspension (*see* **Note 5**). When looking under an inverted light microscope, the membrane vesicles may appear as black dots, which after some incubation time sediment and become immobile as they adhere to the fibroblasts.

4. After incubation, collect the conditioned medium into clean centrifuge tubes.

5. Centrifuge at 15,000×*g* for 15 min at 4 °C and aliquot the supernatant for further analysis. Wash the fibroblasts twice with PBS and lyse the cells with either 500 μl Trizol for RNA measurements or with 200 μl lysis buffer for protein and enzyme analysis (*see* **Note 6**).

3.4 Zymographic Analysis of MMP-2 Secretion by Fibroblasts

These analyses aim at defining the effect of CD147 contained within the membranes by measuring the variation in the expression of potential targets in the fibroblasts (in the example presented here, the variation in the expression of MMP-2). The incubation time of fibroblasts with the membrane vesicles depends on the type of analysis required. For RNA measurements, shorter incubation times (5 min to several hours) will suffice. However, for examining MMP-2 expression in the conditioned medium, usually a 24-h incubation is required to be able to see a clear induction by gelatin zymography. Treatment of fibroblasts with CD147-containing membranes increases MMP-2 secretion from fibroblasts. Therefore, assaying for MMP-2 in the conditioned medium using gelatin zymography is an easy positive control to confirm the activity of CD147 within the preparation of membrane vesicles (Fig. 1).

In the gelatin zymograms, areas of proteolytic activity appear as clear bands against a dark blue background due to proteolytic digestion of the gelatin protein. Both the proenzyme and the active enzyme are seen on the zymogram, since SDS causes activation of the enzyme without causing cleavage of the pro-peptide. Therefore, the molecular weight of the clear bands allows identification as to which is the zymogen or the active enzyme species.

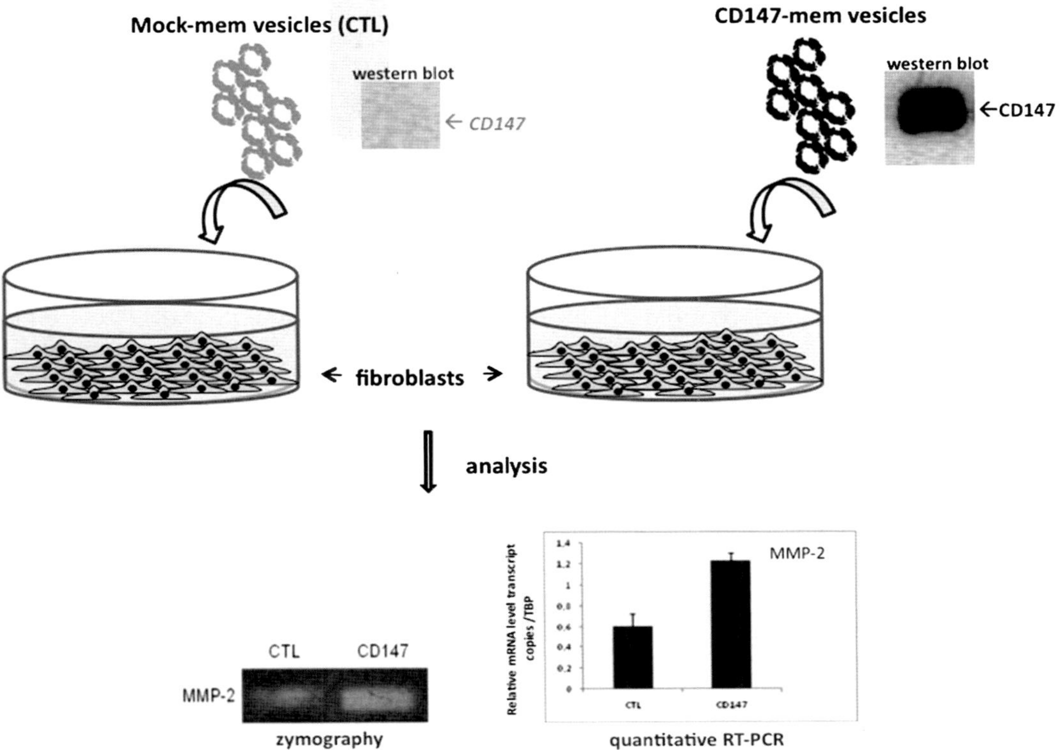

Fig. 1 Increased MMP-2 expression in fibroblasts with CD147 exposure. Example showing that MMP-2 expression is increased when fibroblasts are incubated with membrane vesicles obtained from CD147-transfected CHO cells compared with incubation with mock-transfected CHO membranes, shown by both zymography and quantitative RT-PCR

1. Harvest the serum-free conditioned medium (*see* **Note 7**) after incubation with membranes.

2. Centrifuge at $1{,}500 \times g$ for 15 min, mix one part of this medium with three parts 4× Laemmli sample buffer, and let it stand for 10 min at room temperature. Do not heat.

3. Apply samples (typically 10–25 μl) and run the gel with 1× Tris–glycine SDS running buffer under nonreducing conditions according to the following running conditions: Voltage: 90 V constant, run time: approximately 120 min.

4. Wash gel in 2.5 % Triton X-100 twice for 15 min with gentle agitation to remove SDS and allow renaturation of the enzyme.

5. Replace with Triton X-100 with zymogram developing buffer (approximately 50 ml per gel) and incubate gel at 37 °C for 18–24 h (*see* **Note 8**).

6. Stain with Coomassie Blue solution for 1 h. Gels should be destained with the destaining solution for approximately 2 h, with several changes of the destaining solution until clear bands appear, which represent areas of protease activity.

7. Stained gels can be wrapped in plastic and stored at 4 °C for several months.

This model of direct-ESI mimics cell contact-based communication that may occur in embryogenesis, tumor invasion, and corneal and skin ulcerations. It allows for the examination of the effect on fibroblasts of epithelial membrane proteins present within their natural environment. Although we commonly use corneal fibroblast cell lines, this protocol can be applied to primary fibroblasts or to other stromal cells, such as endothelial cells, depending on the particular interest of the study. This approach may also be compatible with the study of other intrinsic membrane proteins which function via a direct cell–cell contact mechanism and would allow to dissect their biological and molecular effects in such interactions.

4   Notes

1. Acrylamide is highly toxic, so wear gloves when handling.

2. Approximately 150 μg of membrane proteins can be obtained from 1 T75 flask of transfected CHO cells.

3. The T75 flasks containing the transfected CHO cells can be frozen after the PBS wash. They can then be defrosted prior to membrane preparation and use.

4. Different ultracentrifuges can be used according to their availability in the laboratory and different size tubes can be adapted to the volume of the cell lysates. Swing rotors are preferable but if not available, fixed rotors can also be used, in which case the position of the tube in the centrifuge should be marked. This allows positioning of the transparent membrane pellet which can be hard to see, in particular when a small quantity of cells was used.

5. Western blot calibration using purified CD147 has shown that 20 μg of total CD147 transfected CHO membrane proteins contained 0.5 μg of CD147, whereas none could be detected in the mock control membranes.

6. Aliquots can be frozen at –20 or –80 °C awaiting further analysis.

7. Fetal serum contains very high levels of MMP-2; therefore, the incubation of fibroblasts with the membranes should always be done in serum-free medium.

8. Optimal results can be determined empirically by varying the sample load or incubation time.

References

1. Biswas C, Zhang Y, DeCastro R et al (1995) The human tumor cell-derived collagenase stimulatory factor (renamed EMMPRIN) is a member of the immunoglobulin superfamily. Cancer Res 55:434–439
2. Gabison EE, Hoang-Xuan T, Mauviel A et al (2005) EMMPRIN/CD147, an MMP modulator in cancer, development and tissue repair. Biochimie 87:361–368
3. Kataoka H, DeCastro R, Zucker S et al (1993) Tumor cell-derived collagenase-stimulatory factor increases expression of interstitial collagenase, stromelysin, and 72-kDa gelatinase. Cancer Res 53:3154–3158
4. Caudroy S, Polette M, Tournier JM et al (1999) Expression of the extracellular matrix metalloproteinase inducer (EMMPRIN) and the matrix metalloproteinase-2 in bronchopulmonary and breast lesions. J Histochem Cytochem 47:1575–1580
5. Gabison EE, Huet E, Baudouin C et al (2009) Direct epithelial-stromal interaction in corneal wound healing: role of EMMPRIN/CD147 in MMPs induction and beyond. Prog Retin Eye Res 28:19–33
6. Gabison EE, Mourah S, Steinfels E et al (2005) Differential expression of extracellular matrix metalloproteinase inducer (CD147) in normal and ulcerated corneas: role in epithelio-stromal interactions and matrix metalloproteinase induction. Am J Pathol 166:209–219
7. Huet E, Vallee B, Szul D et al (2008) Extracellular matrix metalloproteinase inducer/CD147 promotes myofibroblast differentiation by inducing alpha-smooth muscle actin expression and collagen gel contraction: implications in tissue remodeling. FASEB J 22:1144–1154
8. Miyauchi T, Masuzawa Y, Muramatsu T (1991) The basigin group of the immunoglobulin superfamily: complete conservation of a segment in and around transmembrane domains of human and mouse basigin and chicken HT7 antigen. J Biochem 110:770–774
9. Fadool JM, Linser PJ (1996) Evidence for the formation of multimeric forms of the 5A11/HT7 antigen. Biochem Biophys Res Commun 229:280–286
10. Yoshida S, Shibata M, Yamamoto S et al (2000) Homo-oligomer formation by basigin, an immunoglobulin superfamily member, via its N-terminal immunoglobulin domain. Eur J Biochem 267:4372–4380
11. Bougatef F, Menashi S, Khayati F et al (2010) EMMPRIN promotes melanoma cells malignant properties through a HIF-2alpha mediated up-regulation of VEGF-receptor-2. PLoS One 5:e12265
12. Bougatef F, Quemener C, Kellouche S et al (2009) EMMPRIN promotes angiogenesis through hypoxia-inducible factor-2alpha-mediated regulation of soluble VEGF isoforms and their receptor VEGFR-2. Blood 114:5547–5556
13. Vishwanath M, Ma L, Otey CA et al (2003) Modulation of corneal fibroblast contractility within fibrillar collagen matrices. Invest Ophthalmol Vis Sci 44:4724–4735

Chapter 10

Microencapsulation of Stem Cells to Study Cellular Interactions

Keith Moore, Adam Vandergriff, and Jay D. Potts

Abstract

Microencapsulation is a technique used in both controlled delivery of materials over time as well as preservation of these materials while delivery is occurring. The range of materials able to be encapsulated is variable, from drugs to living cells. The latter is described here. Electrospray microencapsulation applies a high-voltage field, through which a polymeric material is extruded. A gelling bath, comprising a cross-linking material, is used to create a stable hydrogel containing secondary substances intended for delivery. Control of extrusion parameters, such as flow rate and voltage, allows for specification of diameter and pore sizes of the microcapsules.

Key words Electrospray, Microencapsulation, Bone marrow stromal cells, Sodium alginate, Transfection, Electrohydrodynamics

1   Introduction

Microencapsulation is a technique in which small particles are coated with a polymeric material to create a delivery system with secondary properties. The secondary property is controlled release or protection of the encapsulated material from potentially harmful environments during delivery. Examples include areas of damaging pH or toxic environments [1–4]. Many methods exist for microencapsulation including pan coating, electrospray, emulsion, and pulse-flow. Furthermore, within the classification of electrospray microencapsulation exist multiple techniques to carry out synthesis. Here the electrospray method is presented in a one-step method using a single syringe. Each delivery system has inherent properties that make it ideal for individual applications. When encapsulating cells, biocompatibility of the synthesis materials is an important factor. Sodium alginate is a polymeric material derived from brown algae and is nontoxic to cell survival [5, 6]. When cross-linked to cations, such as calcium, a hydrogel is formed [7, 8], making this polymer an

Troy A. Baudino (ed.), *Cell-Cell Interactions: Methods and Protocols*, Methods in Molecular Biology, vol. 1066,
DOI 10.1007/978-1-62703-604-7_10, © Springer Science+Business Media New York 2013

ideal choice. Other polymers, such as polyethylene glycol (PEG) and chitosan, have been shown to provide biocompatibility as well.

The cell line used here is bone marrow stromal cells (BMSCs), also known as mesenchymal stem cells. These mesenchymal cells are ideal progenitor cells for regenerative medicine due to their ability to differentiate into adiopocytes, osteoblasts, chondrocytes, and neural and endothelial cells [9, 10]. In addition, BMSCs provide an ideal cell based on the ease of which they are isolated from extracted bone marrow of multiple species including rats, mice, chickens, felines, and humans [11, 12]. Microencapsulation of drugs often requires modification of the capsule properties due to rapid diffusion and burst release of the encapsulated materials immediately after synthesis [13]. As a way to provide continuous application, cells may be transfected to overexpress potentially therapeutic proteins. These transfected cells are then microencapsulated to deliver to a specific site and release these cells and secreted proteins that the cell is synthesizing. Targeted delivery of these cells and products have been widely attempted to study such injuries as myocardial infarcts, ocular insults, and skin wounds. The protocol here provides a basis to isolate, transfect as desired, and microencapsulate BMSCs for targeted delivery.

2 Materials

All materials should be sterile to prevent contamination of cells. Solutions containing water should be made with ultrapure, filtered, 18 MΩ water that was autoclaved before use. All materials must be heated to 37 °C prior to use with the stem cells.

2.1 Isolation and Culture of Bone Marrow Stromal Cells

1. Penicillin–streptomycin stock solution for medium: 100 μg/ml streptomycin and 100 units/ml penicillin G.

2. Primary medium (500 ml): 2.25 g Dulbecco's modified Eagle's medium (DMEM), 0.6 g sodium bicarbonate, 1.787 g HEPES. Adjust pH to 7.4, add 10 % fetal bovine serum (FBS), 50 μg/ml gentamicin, 5 ml Pen–Strep stock solution, and 1 ml amphocyterin B.

3. Tibia/femur medium (500 ml): 2.25 g DMEM, 0.6 g sodium bicarbonate, 1.787 g HEPES. Adjust pH to 7.4 and then add 10 % FBS and 125 μg/mL gentamicin.

4. Moscona's solution (500 ml): 4.0 g NaCl, 0.1 g KCl, 0.5 g NaHCO$_3$, 0.85 g glucose, and 0.0025 g NaH$_2$PO$_4$ H$_2$O. Adjust pH to 7.4.

5. 0.25 % Trypsin–EDTA (1L): 2.5 g Trypsin in 500 ml Moscona's solution. Dissolve 1 g EDTA in another 500 ml of Moscona's solution with heat for 5–10 min. Mix both solutions together, adjust pH to 7.4, filter into 50 ml conical tubes, and store frozen at –20 °C until used.

6. Rats.

7. 18- and 21-G needles.

8. 25 ml syringes.

9. 15 and 50 ml conical tubes.

10. Bone scissors.

11. T75 tissue culture flasks.

12. Lipofectamine (Invitrogen, Grand Island, NY, USA).

2.2 Synthesis of Sodium Alginate Microcapsules

1. 2 % Sodium alginate: For 5 ml, add 0.1 g sodium alginate to 5 ml of water in a sterile 15 ml conical tube. Heat for 15–20 min at 37 °C until dissolved (*see* **Note 1**). Filter using a 0.2 μm syringe filter with a cellulose acetate membrane.

2. HEPES buffer: Add 119.15 g HEPES to 500 ml of water. Adjust pH to 7.4 and store in the dark.

3. 0.15 M calcium chloride ($CaCl_2$) gelling bath (500 ml): Add 11.025 g $CaCl_2$ to 500 ml of water and adjust pH to 7.4 using HEPES buffer.

4. 3 ml syringes.

5. 30-G blunt needles.

6. 50 ml beakers.

7. Voltage generator.

3 Methods

3.1 Cell Isolation

1. Humanely sacrifice the rat using preferred method (*see* **Note 2**). This should be performed according to individual institutional policy. Some examples are:

   (a) Carbon dioxide asphyxiation.

   (b) Overdose of an anesthesia, such as isoflurane.

2. Make an incision from the medial ankle to the base of the pelvis using a scalpel, cutting through the skin to reveal the underlying muscle.

3. Make a circumferential incision around the ankle.

4. Peel the skin back from the ankle towards the pelvis using a scalpel or a scissors to remove any connective tissue.

5. Cut back and remove any muscle around the medial area of the hip joint.

6. Using bone scissors, cut the femur as close as possible to the hip joint. This should be done carefully to prevent puncture of internal organs, creating a release of large amounts of blood.

7. Remove the leg from the body by cutting any remaining connective tissue or muscle.

8. Using bone scissors, cut the ankle below the tibia to completely remove the foot.

9. Remove any remaining muscle tissue from the leg so that the tibia and femur are completely exposed (*see* **Note 3**).

10. Use bone scissors to cut through the knee joint by flexing the knee around the scissor blades prior to cutting. The joint can be felt by placing your index finger on the knee prior to cutting.

11. Place the femur in a Petri dish filled with approximately 10 ml of tibia/femur medium.

12. Remove and discard the fibula from tibia. Place tibia in the dish of tibia/femur medium with the femur. Discard the fibula.

13. Fill a 50 ml conical tube with primary culture medium.

14. Using an 18-G needle, fill a 25 ml syringe with primary culture media from the conical tube.

15. Cut the ends of the femur off using bone scissors, creating a tube from which the marrow can be extracted.

16. Place the femur over a second conical tube, insert the syringe/needle of primary media, and flush out the marrow. Do so from both ends of the bone, moving the needle around, until the bone appears translucent.

17. Repeat this process with the tibia using a 21-G needle, flushing the marrow into the conical tube containing the marrow from the femur.

18. Spin down the medium/marrow mixture for 8 min at approximately $125 \times g$.

19. Aspirate the supernatant from the bone marrow pellet.

20. Suspend the marrow/cells in 15 ml of primary culture medium and transfer to a T75 culture flask. Incubate at 37 °C 5 % CO_2.

21. Do not remove the medium for at least 1 week, checking the cell confluency daily.

22. After 1 week, or when cells reach at least 90 % confluency, passage the cells (*see* **Note 4**).

3.2 Cell Passaging

All materials used in cell culture should be sterile and warmed to 37 °C before use. Perform all procedures in a sterile cell culture hood.

1. Aspirate the primary medium from the 90–95 % confluent T75 flask.

2. Rinse the cells with 10 ml of Moscona's solution to remove any excess media. Aspirate the Moscona's from the flask.

3. Add 1.5 ml trypsin/EDTA to the flask. Immediately rock the flask for 2–3 min until the cells are no longer attached. This may be confirmed microscopically.

4. Add 10 ml primary medium to neutralize the trypsin/EDTA.

5. Transfer the cells and medium to a 50 ml conical tube and centrifuge for 8 min at $125 \times g$.

6. Aspirate the supernatant from the cell pellet and resuspend in 10 ml of primary culture medium. Transfer to two new T75 flasks, with 5 ml medium/cells going into each flask.

7. Add additional medium to each flask to reach a total volume of 15 ml.

8. Place in incubator at 37 °C 5 % CO_2.

9. Allow 1–2 days for cells to attach before replacing media.

3.3 Transfection of BMSCs

Transfection of BMSCs is carried out according to the Invitrogen Lipofectamine 2000 protocol for transfecting with plasmid DNA. All compatible solutions were purchased according to the protocol guidelines and should be followed as published there.

1. Culture cells to 90–95 % confluence in 60 mm tissue culture dishes. Other sizes may be substituted as desired.

2. Follow Invitrogen protocol for addition of DNA/Lipofectamine 2000.

3. Once cells are transfected a sample dish should be tested to ensure efficiency. This can be done by lysing the cells and running the protein on a 7 % SDS-PAGE gel according to individual lab's western blotting protocol.

3.4 Electrospray Microencapsulation of Cells

All actions should be performed in a sterile cell culture hood at room temperature. All solutions should be at 37 °C. All microencapsulation components should be sterilized prior to use if coming in direct contact with solutions/materials/cells.

1. Add 1.5 ml of 0.25 % trypsin–EDTA to a 90–95 % confluent T75 flask of transfected BMSCs to remove the cells as described in the cell culture protocol (Method 3.2.). Transfer to a 15 ml vial, centrifuge, and resuspend in 1 ml of primary medium.

2. Count the cells using a hemocytometer by making a 1:10 dilution of Erythrosin Blue Dye (0.05 % w/v) to cells in suspension (*see* **Note 5**).

3. Once the cell count is obtained, calculate the volume, in microliters, needed to create a solution that is 500,000 cells/ml. Pipette this volume of cells in suspension into a 1.5 ml microcentrifuge tube and gently centrifuge at $125 \times g$ for 2–3 min.

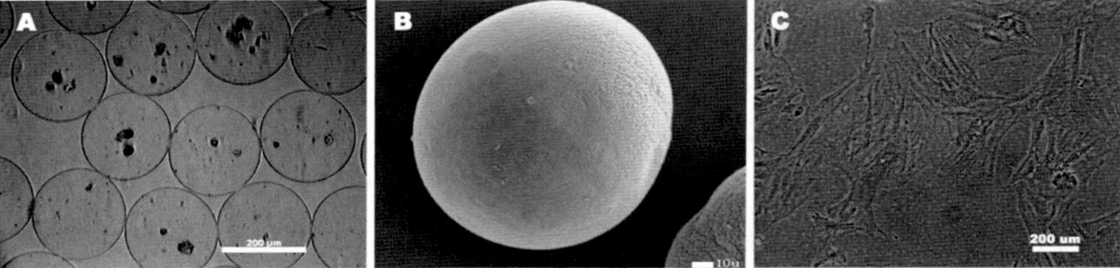

Fig. 1 Encapsulated BMS cells and components. (**a**) BMSCs encapsulated in alginate 200 μm microcapsules. On average there are 3–10 cells per microcapsule. This can be manipulated with varying the concentration of cells in the alginate solution. (**b**) An SEM image of the microcapsule illustrating the spherical nature of the capsule. (**c**) BMSCs shown in culture. These cells take on a stellate morphology and depending on the confluency appear similar to fibroblasts in culture

4. Remove the excess medium and add 250 μL of 2 % sodium alginate. Gently mix by pipetting several times (*see* **Note 6**).

5. Add the cell/alginate mixture to a 3 ml syringe, making sure to remove any air bubbles. Attach a stainless steel blunt-tip needle (30 G), and then place on a syringe driver or the flow rate control device being used.

6. Set the syringe pump to 30 mm/h and connect to a ring stand (*see* **Note 7**).

7. Place 30 ml of 0.15 M $CaCl_2$ gelling solution in a 50 ml beaker. Move the beaker so that the syringe driver/needle sits centered directly above (*see* **Note 8**).

8. Attach the positive cable of the voltage generator to the needle tip, making sure that there is a tight connection. Place a ground in the gelling solution. This can be done by attaching a small piece of copper wire to the beaker in such a way that it partially sits in the solution and partially hangs over the beaker lip. Attach the negative cable to the exposed copper wire.

9. Start the syringe pump to begin extrusion of the cell/alginate mixture into the gelling bath. Immediately after starting the syringe pump, the voltage generator should be turned on and set to a voltage of 6.0 kV for 200 μm diameter microcapsules (*see* **Note 9**).

10. Once the cell/alginate mix has been completely extruded into the gelling bath, allow the microcapsules to cross-link for 8 min. Remove excess gelling solution, place the microcapsules in a Petri dish, and cover in primary medium (Fig. 1).

11. Incubate at 37 °C in 5 % CO_2 till using in further studies.

4 Notes

1. Add alginate to water, not water to alginate. This prevents a difficult-to-dissolve pellet from forming at the bottom of the tube. A combination of heating and then vortexing several times will dissolve the alginate. Initially the solution is cloudy. When the alginate is ready the solution will be clear/slightly opaque.

2. Rat type and age can be varied. Typically male Sprague Dawley rats around an approximate age of 1 month have been used in our BMSC isolations.

3. Sterile tissue is often helpful to pull the remaining tissue off of the bone rather than attempting to cut this away.

4. Typically BMSCs are not used experimentally until passage 3–4. If desired these cells may be microencapsulated at additional passages, based on individual experimental design.

5. Any cell counting dye or lab-specific protocol may be substituted here. If using the Erythrosin Blue protocol a dilution of 450 µl of dye to 50 µl cells in suspension is sufficient for counting.

6. The ideal cell number is between 500,000 and 1,000,000 cells/ml for BMSCs. A higher amount leads to lower cell viability once encapsulated. Other cell types may require higher or lower amounts based on cell diameter. When preparing this stage take care not to shear the cells while mixing.

7. Flow rate is variable based on the device being used. A high flow rate will shear the cells and should be avoided.

8. The needle to working distance is an essential factor in extruding the alginate through a sufficient voltage field. A distance from needle tip to solution of 7 mm has been found to be sufficient. Depending on the beaker size, the volume listed here may be adjusted to meet this distance.

9. It is essential that little to no current flow through the cells. The voltage settings may vary from instrument to instrument and can be adjusted to change the microcapsule diameter. The voltage listed here produces microcapsules with an average diameter of approximately 200 µm on the tested instrument.

Acknowledgements

This work was supported by NIH grant HL086901 (JDP).

References

1. Chew S, Wen Y, Dzenis Y et al (2006) The role of electrospinning in the emerging field of nanomedicine. Curr Pharm Des 12:4751–4770
2. Pham Q, Sharma U, Mikos A (2006) Electrospinning of polymeric nanofibers for tissue engineering applications: a review. Tissue Eng 12:1197–1211
3. Rutledge G, Fridrikh S (2007) Formation of fibers by electrospinning. Adv Drug Deliv Rev 59:1384–1391
4. Chakraborty S, Liao I, Adler A et al (2009) Electrohydrodynamics: a facile technique to fabricate drug delivery systems. Adv Drug Deliv Rev 61:1043–1054
5. Lin J, Yu W, Liu X et al (2008) In vitro and in vivo characterization of alginate-chitosan-alginate artificial microcapsules for therapeutic oral delivery of live bacterial cells. J Biosci Bioeng 105:660–665
6. Zhang W, Yang G, Zhang A et al (2010) Preferential vitrification of water in small alginate microcapsules significantly augments cell cryopreservation by vitrification. Biomed Microdevices 12:89–96
7. Paul A, Shum-Tim D, Prakash S (2010) Investigation on PEG integrated alginate-chitosan microcapsules for myocardial therapy using marrow stem cells genetically modified by recombinant baculovirus. Cardiovasc Eng Technol 1:154–164
8. Zhang WJ, Li BG, Zhang C et al (2008) Biocompatibility and membrane strength of C3H10T1/2 cell-loaded alginate-based microcapsules. Cryotherapy 10:90–97
9. Gronthos S, Akintoye S, Wang CY et al (2006) Bone marrow stromal stem cells for tissue engineering. Periodontology 2000(41):188–195
10. Jiang Y, Jahagirdar B, Reinhardt RL et al (2002) Pluripotency of mesenchymal stem cells derived from adult marrow. Nature 418:41–49
11. Khatri M, O'Brien T, Sharma JM (2009) Isolation and differentiation of chicken mesenchymal stem cells from bone marrow. Stem Cells Dev 18:1485–1492
12. Martin D, Cox N, Hathcock TL et al (2002) Isolation and characterization of multipotential mesenchymal stem cells from feline bone marrow. Exp Hematol 30:879–886
13. Moore K, Amos J, Davis J et al (2012) Characterization of polymeric microcapsules containing a low molecular weight peptide for controlled release. Microsc Microanal 19:213–226

Chapter 11

Cell-Surface Protein–Protein Interaction Analysis with Time-Resolved FRET and Snap-Tag Technologies

Timothy N. Feinstein

Abstract

Förster resonance energy transfer (FRET) is a proximity-dependent quantum effect that allows the measurement of protein interactions and conformational changes which are invisible to traditional forms of fluorescence or electron microscopy. However, FRET experiments often have difficulty detecting interactions that are transient and localized or occur in low abundance against a large background. This protocol describes a method of improving on the sensitivity and quantifiability of FRET experiments by using time-specific detection to isolate FRET-mediated acceptor emission from cross-talk excitation and all other sources of nonspecific fluorescence background.

Key words Förster resonance energy transfer, FRET, Protein–protein interactions, SNAP tagging, G-protein-coupled receptors, GPCR

---

1 Introduction

Whereas one would normally use biochemistry to study protein–protein interactions and conformational changes, such as co-immunoprecipitation, western blotting, GST pull-down and analytical ultracentrifugation, numerous advances in optical microscopy now allow us to measure these phenomena in real time and in living cells or even intact tissue. The key to these developments is Förster resonance energy transfer (FRET), a quantum effect wherein energy from an excited fluorescent "donor" molecule jumps a small spatial gap to an "acceptor" molecule that absorbs light at about the same wavelength that the "donor" molecule emits when excited. Emission of light from molecules excited by intermolecular FRET [1] is a better measure of two molecules close together than conventional microscopy can achieve, even taking into account recent advances that extend spatial resolution beyond traditional diffraction-imposed limits [2]. The efficiency of energy transfer drops with the sixth power of distance and thus Förster radii, the distance of 50 % energy transfer, tend to be quite

Troy A. Baudino (ed.), *Cell-Cell Interactions: Methods and Protocols*, Methods in Molecular Biology, vol. 1066, DOI 10.1007/978-1-62703-604-7_11, © Springer Science+Business Media New York 2013

small (50–110 Å for most FRET pairs [3]) and tightly constrained. Significant FRET between two proteins of interest (POIs) thus shows that the two proteins almost certainly either bind one another or else belong to a common protein complex. Quantum transfer parameters can be determined for each FRET pair through analysis with a spectrofluorometer, and this allows for the quantitative estimation of the average distance between molecules within a cell, measured area or volume after applying the appropriate corrections [4].

However, next to biochemical interaction tests such as co-immunoprecipitation the measurement of donor emission intensity has some disadvantages. Most FRET donors emit light in a broad intensity range, to a degree that light detected using filter sets for acceptor emission includes a nontrivial amount of light from donor emission as well ("bleed-through") and some fraction of acceptor molecules will be excited by light meant to excite the donor ("cross talk"). As a result, weak, infrequent, or transient interactions that are detectable by western blotting can be difficult or impossible to detect with FRET due to background noise from bleed-through and cross talk. This apparent puzzle can be overcome by an intriguing property of light that arises from energy transfer: both donor and acceptor emission from excitation by cross-talk emit light that decays rapidly and in a mono-exponential fashion after a short pulse of excitation light, whereas FRET-mediated light emission decays more slowly [5]. If one excites the acceptor with a short enough pulse of light and measures emission from the FRET acceptor after a carefully timed delay, it is possible to detect and quantify a small FRET signal amidst an overwhelming background of bleed-through and cross talk. Initial demonstrations with conventional FRET pairs required expensive, specialized equipment to exploit emission delays on the scale of a few nanoseconds [5]. However, the use of more exotic FRET pairs that have much longer emission lifetimes allows the detection of time-resolved FRET (TR-FRET) emission with equipment that is much simpler and less expensive. The most appealing of these new FRET donors are the lanthanide elements 62 and 64, europium and terbium, which emit for up to 100 µs after excitation [6]. Europium emission lasts three orders of magnitude longer than the decay time of matched acceptors such as fluorescein/Alexa488 and Cy5/Alexa647, ruling out the possibility of cross talk, and bleed-through is eliminated by a convenient lack of overlap between light emitted by europium/terbium and the emission of most far-red acceptor proteins (Fig. 1). This delay is exploitable by the more conventional optics used in plate reader-based screens for time-resolved FRET. This approach is known as homogenous TR-FRET, or HTRF, because the plate reader averages the emission of all proteins in a cross-section of cells within the well being analyzed.

Living cells have no natural means to take up lanthanide elements or incorporate them into a transgenic POI. To perform

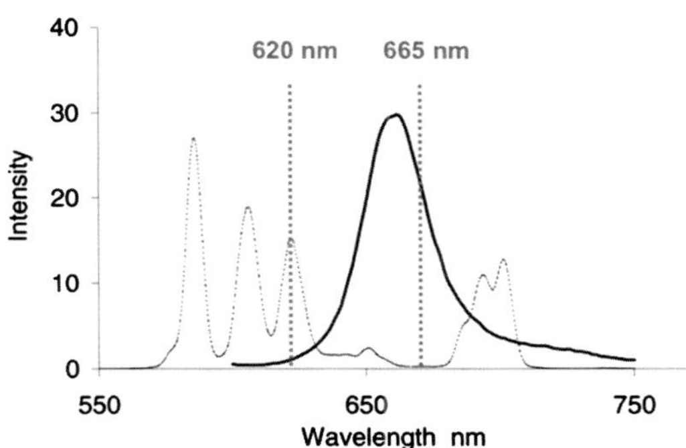

Fig. 1 Spectral selectivity of HTRF assays. The spectral selectivity of HTRF assays is shown by overlaying the emission spectrum of europium cryptate (*dotted line*), measured at 620 nm, with the emission of a far-red acceptor (*solid line*), measured at 665 nm

HTRF, it is therefore necessary to either covalently label a purified protein, peptide, or small molecule that binds an extracellular domain of your POI [7] or use an antibody or genetically coded custom-inducible covalent linkage to specifically attach the donor/acceptor molecules to your POI [8]. A number of genetically encodable motifs have been developed through the modification of proteins that repair DNA, bridge peptide backbones, or add covalent modifications to a specific protein motif [9]. An appealing example is the SNAP tag, a 20 kDa mutant of the O6-alkylguanine-DNA alkyltransferase that binds and makes a covalent linkage with belzylguanine (BG) derivatives, allowing the simple and irreversible labeling of a SNAP-modified POI with any fluorophore that has been conjugated with a BG moiety [10]. A general protocol for HTRF analysis of cell-surface protein–protein interactions using SNAP tags and BG-derivatized FRET pairs is described here.

2   Materials

2.1   Cell Culture Reagents

1. Culture medium: For HEK293 cells, Dulbecco's modified Eagle's medium (DMEM) supplemented with 10 % fetal calf serum (FCS) and 1 U/ml penicillin/streptomycin (optional).

2. Phosphate-buffered saline (PBS), pH 7.4.

3. 0.05 % Trypsin/0.2 mg/ml EDTA in PBS (trypsin–EDTA).

4. Hemacytometer.

5. Imaging medium: Tris-Krebs buffer (20 mM Tris–HCl, 118 mM NaCl, 5.6 mM glucose, 1.2 mM KH_2PO_4, 1.2 mM $MgSO_4$, 4.7 mM KCl, 1.8 mM $CaCl_2$, pH 7.4) supplemented with 0.1 % (w/v) BSA and 0.1 % (w/v) glucose.

6. Amaxa Nucleofector I (Lonza, Basel, Switzerland).

7. Nucleofector Kit V (Lonza).

2.2 Plasmid Cloning	Methods and reagents for adding epitope tags to cloned GPCRs [11] and site-directed mutagenesis [12] are described in other editions of *Methods in Molecular Biology* (*see* **Note 1**).

2.3 SNAP-Tagging Reagents

1. Donor molecules: Benzylguanine-bound europium cryptate (e.g., BG-TBP; Covalys Biosciences, Witterswil, Switzerland) or terbium cryptate (e.g., SNAP-Lumi4-Tb; Cisbio, Bedford, MA, USA).

2. Acceptor molecules: Benzylguanine-tagged fluorophores (e.g., SNAP-Surface 632 or SNAP-Surface Alexa Fluor 488; New England Biolabs, Ipswich, MA, USA; or BG-647; Covalys).

3. Custom synthesis of benzylguanine-labeled fluorophores is described [8].

2.4 ELISA

1. Primary antibodies: HA.11 Anti-HA monoclonal antibody (Covance, Princeton, NJ, USA); anti-FLAG polyclonal antibody (Sigma-Aldrich, St. Louis, MO, USA).

2. Anti-mouse and anti-rabbit secondary antibodies labeled with horseradish peroxidase (HRP).

3. HRP substrate (e.g., OPD; Thermo Scientific).

4. Controls: Purified HA- or FLAG-tagged protein (e.g., Alpha Diagnostic Intl., San Antonio, TX, USA; #HA15-R, FLAG15-R).

5. HTRF-capable plate reader (e.g.: Rubystar; BMG Labtechnologies, Cary, NC, USA).

6. Black multiwell plates (e.g., PerkinElmer #6005279).

---

3 Methods

3.1 Prepare Donor and Acceptor Constructs

Before cloning SNAP- and HA/FLAG-tagged POIs, verify that each is a transmembrane protein that is expressed to the plasma membrane and has a free extracellular N- or C-terminus that can be modified without inhibiting function of the protein (*see* **Note 2**). When possible, SNAP-tagged receptor proteins should be tested for activity using pharmacological assays such as radioligand binding [13] and/or generation of second messengers, such as calcium or cAMP [14]. Correct plasma membrane localization should be verified by immunocytochemistry [15].

3.2 Transfection

A protocol for transfection of mammalian cells by electroporation is described generally in ref. 16. You can also perform transfections as per individual lab protocols.

1. For transfection of HEK293 cells, passage cells in a 25 cm² flask and culture for 2–3 days, until 80–90 % confluent. Using cells

that are grown at a greater density, or cells that have been passaged more than 20 times, will reduce transfection efficiency.

2. Add trypsin–EDTA, incubate for 5 min or until cells have detached from the flask, and then neutralize the trypsin by adding 5 ml fresh medium.

3. Count the cells using a hemacytometer.

4. Centrifuge cells at $200 \times g$ for 10 min and resuspend at 1×10^6 cells in 100 µl Nucleofector solution with supplement.

5. Add 3 µg of plasmid DNA, transfer to a suitable electroporation cuvette, and select Nucleofector program Q-001.

6. After electroporation, immediately add 500–800 µl of warm culture medium and divide the cells between the wells of a black multiwell plate. Each well should already contain medium warmed and equilibrated in a 37 °C CO_2 incubator (*see* **Note 3**). This should be performed in duplicate, as one plate will be used for experiments while the other will be trypsinized and cells counted to determine the precise cell number per well at the time of experiment.

3.3 Validate the Expression of Proteins

1. To measure the number of POI per cell, plate a known quantity of cells on a multiwell dish, such as 5×10^5 HEK293 cells per well in a 24-well dish.

2. After 24 h, trypsinize three of the wells and count the cells using a hemacytometer.

3. Perform an ELISA as previously described [17] using a standard curve of adsorbed HA- and FLAG-tagged control proteins to estimate POI/cell (*see* **Note 4**).

4. To verify the expression of each POI in co-transfected cells, perform an ELISA as described in side-by-side triplicates using three wells for each primary antibody (six total) (*see* **Note 5**).

3.4 Labeling Protocol

1. SNAP tagging: To label POIs expressing the SNAP-tag motif on their extracellular domain, incubate cells growing on black multiwell plates with BG-labeled fluorophores for 1 h in a humidified incubator at 5 % CO_2 and 37 °C. The absolute and relative concentration of BG-labeled fluorophores will depend on experimental needs. Wash the cells 4× in Tris-Krebs buffer and image.

2. Antibody labeling: Add BG-donor along with anti-FLAG or anti-HA antibody (2 nM) and incubate overnight (*see* **Note 6**).

3.5 Detecting Homogenous Time-Resolved FRET

The exact method of HTRF detection will depend on the specifications of the specific HTRF-equipped plate reader. When detecting homo- and heterodimerization of $GABA_B$ receptors using a Rubystar plate reader [8], these settings were used: Excitation

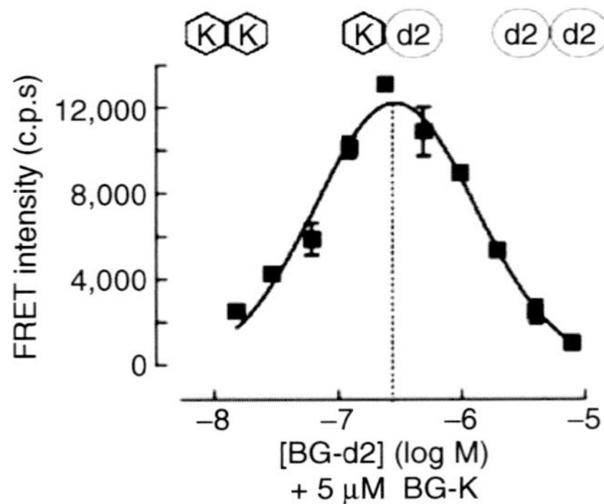

Fig. 2 Demonstration of the dimeric nature of GABA$_B$-receptor binding. Rat ABA$_{B1}$ and GABA$_{B2}$ receptors, tagged at their N-termini with the SNAP-tag motif, were incubated with varying concentrations of donor and acceptor, and FRET intensity was measured by HTRF. The sharp peak at roughly equimolar donor:acceptor concentrations shows that GABA$_B$ receptors are predominantly found in dimeric complexes. Reprinted from [8] with permission from Nature Publishing Group

wavelengths: Europium cryptate, 337 nm; far-red acceptor (Alexa 647, Cy5, DY-647, SNAP-Surface 632): 620–650 nm. Fluorescence detection: 620 nm (donor) and 682 nm (acceptor). Time-resolved detection delay: 50 ms. Agonist/antagonist challenge, drug addition, and washing with fresh medium can be done using optional perfusion modules or by removing the plate and manually adding reagents or washing between measurements.

3.6 Specificity of Labeling and FRET

To verify the specificity of BG-SNAP binding and to ensure that unbound fluorophores have been thoroughly washed out, it is important to correlate the amount of bound donor or acceptor dye, as measured by relative fluorescence or ELISA with the average number of receptors per cell. In the case of receptor proteins such as G-protein-coupled receptors, POI numbers can be estimated through radioligand binding.

A key test of FRET specificity is to measure its sensitivity to changes in the ratio of donor to acceptor molecules. Specific FRET will show a hyperbolic response to increasing relative concentrations of the donor, saturating at a point above 1:1 parity that depends on the affinity and extent of the interaction in question [18] (*see* **Note 7**). When detecting homodimers using identical POIs labeled with donor/acceptor SNAP conjugates, specific FRET will follow a negative parabolic response, peaking at 1:1 parity and declining at excessive relative concentrations of either donor or acceptor (Fig. 2) (*see* **Note 8**). Nonspecific FRET increases in a linear or exponential fashion with the total concentration of acceptor and does not saturate (Fig. 3) [19, 20]. When measuring homodimers, altering the ratio of donor:acceptor is accomplished by manipulating the concentration of each BG-conjugated

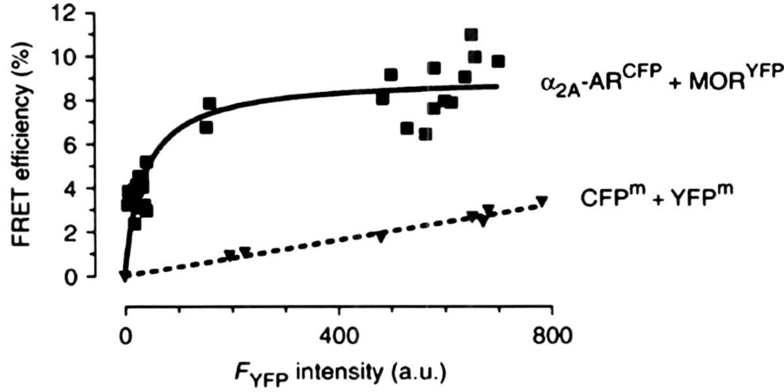

Fig. 3 Experimental demonstration of FRET specificity. HEK293 cells were transfected with plasmids encoding either the a_{2A}-adrenergic receptor and the m opioid receptor (MOR) tagged at their C-termini with CFP and YFP (*solid line*) or else modified CFP and YFP proteins that localize to the plasma membrane (CFP^m/YFP^m, *dotted line*). Titrated expression of the YFP acceptor molecule was used to demonstrate the saturability of FRET between a_{2A}-AR^{CFP} and MOR^{YFP}, whereas FRET between CFP^m and YFP^m was weak and showed a linear dependence on acceptor concentration. Reprinted from [23] with permission from Nature Publishing Group

fluorophore. In heterogenous FRET, the donor: acceptor ratio can be manipulated by changing the relative amount of each plasmid added during transfection. Experiments should include parallel, negative control trials using acceptor POIs that are not expected to bind one another.

A quantitative technique, FRET allows measurement of the average distance between molecules, after granting certain assumptions, or the average efficiency of energy transfer. For a thorough discussion of how these measurements are done and can be used to estimate biophysical parameters, *see* ref. 21.

4 Notes

1. In the case of GPCRs, as well as other cell surface receptors, modifications to the N- or the C-terminus can have significant consequences for protein expression, function, and localization to the plasma membrane. For this reason it is important to perform preliminary tests such as those described in Subheading 3.1, as well as any other test for functions that may be important to the specific activity being assayed. The time spent on due diligence is worth more than the time spent backtracking and correcting after a new construct is found to be problematic. DNA sequencing of the new expression construct is also an important form of quality control to ensure that stop codons and frameshift errors are not erroneously added.

2. Proteins tagged with the SNAP motif at their N-terminus are somewhat more stable than proteins tagged at the C-terminus, so choose this site when possible.

3. For optimal cell viability, keep cells in Nucleofector solution for as short a time as possible and resuspend cells in fresh medium immediately after electroporation. The amount of plasmid DNA used can vary between 1 and 5 μg, with higher concentrations increasing expression but generally decreasing cell viability. Each plasmid combination should be titrated to find the optimal parameters.

4. Also using the ELISA kit, it is important to perform preliminary experiments to determine when peak transgene expression occurs and how long it lasts. To do this, perform one transfection and plate six wells per transfection on each of the five transparent-bottom 96-well plates. On each day following transfection, count cells from three of the six wells with a hemacytometer and test expression in the other three by ELISA to determine protein expression per cell.

5. An equal expression of donor and acceptor is ideal in terms of POI number per cell. The relative amount of each plasmid transfected should be titrated until near-equal representation is achieved. Note that protein stability at the plasma membrane is not equal, so a plasmid ratio that yields equal representation on day 2 after transfection may not remain equal on day 3 or 4.

6. An alternative way to measure FRET between two distinct proteins is the CLIP tagging protocol, which uses a moiety made from the human O6-alkylguanine-DNA alkyltransferase (hAGT) to covalently bind labels through benzylcytosine (BC). SNAP-CLIP multiplexing can potentially provide the same benefits as antibody-SNAP FRET with a shorter incubation, greater labeling specificity, and less risk of antibody-mediated cross-linking [22]. Other alternatives include acyl carrier protein (ACP) tagging and a modification (MCP), which bind coenzyme A (CoA) derivatives [9].

7. If proteins form dimers and not higher oligomers, then the FRET intensity curve will have a peak width at half maximum (PWHM) of 1–1.5 log units and will approach zero at disparities greater than 1,000× in either direction. Higher order oligomeric complexes will exhibit a wider peak that approaches zero at much greater concentration disparities.

8. When measuring homodimerization in this way, the peak concentration-dependent efficiency of energy transfer should be independent of the concentration of SNAP-tagged receptor

present. This can be verified by side-by-side ELISA and HTRF analysis of different transfection regimes. Note that this approach is only valid if there are no native or unlabeled potential binding partners expressed on the plasma membrane.

References

1. Selvin PR (2000) The renaissance of fluorescence resonance energy transfer. Nat Struct Biol 7:730–734
2. Gould TJ, Hess ST, Bewersdorf J (2012) Optical nanoscopy: from acquisition to analysis. Annu Rev Biomed Eng 14:231–254
3. Lam AJ, St-Pierre F, Gong Y et al (2012) Improving FRET dynamic range with bright green and red fluorescent proteins. Nat Methods 9:1005–1012
4. Xia Z, Liu Y (2001) Reliable and global measurement of fluorescence resonance energy transfer using fluorescence microscopes. Biophys J 81:2395–2402
5. Wallrabe H, Periasamy A (2005) Imaging protein molecules using FRET and FLIM microscopy. Curr Opin Biotechnol 16: 19–27
6. Bazin H, Trinquet E, Mathis G (2002) Time resolved amplification of cryptate emission: a versatile technology to trace biomolecular interactions. J Biotechnol 82:233–250
7. Albizu L, Cottet M, Kralikova M et al (2010) Time-resolved FRET between GPCR ligands reveals oligomers in native tissues. Nat Chem Biol 6:587–594
8. Maurel D, Comps-Agrar L, Brock C et al (2008) Cell-surface protein–protein interaction analysis with time-resolved FRET and snap-tag technologies: application to GPCR oligomerization. Nat Methods 5:561–567
9. Yano Y, Matsuzaki K (2009) Tag-probe labeling methods for live-cell imaging of membrane proteins. Biochim Biophys Acta 1788: 2124–2131
10. Keppler A, Gendreizig S, Gronemeyer T et al (2003) A general method for the covalent labeling of fusion proteins with small molecules *in vivo*. Nat Biotechnol 21:86–89
11. Huang Y, Willars GB (2011) Generation of epitope-tagged GPCRs. Methods Mol Biol 746:53–84
12. Braman J, Papworth C, Greener A (1996) Site-directed mutagenesis using double-stranded plasmid DNA templates. Methods Mol Biol 57:31–44
13. Bylund DB, Deupree JD, Toews ML (2004) Radioligand-binding methods for membrane preparations and intact cells. Methods Mol Biol 259:1–28
14. Hoffmann C, Gaietta G, Bunemann M et al (2005) A FlAsH-based FRET approach to determine G protein-coupled receptors: the quest for functionally selective conformations is open. Nat Methods 2:171–176
15. Petralia RS, Wenthold RJ (1999) Immunocytochemistry of NMDA receptors. Methods Mol Biol 128:73–92
16. Lakshmipathy U, Buckley S, Verfaillie C (2007) Gene transfer Via nucleofectin into adult and embryonic stem cells. Methods Mol Biol 407:115–126
17. Gaastra W (1984) Enzyme-linked immunosorbent assay (ELISA). Methods Mol Biol 1:349–355
18. Tramier M, Piolot T, Gautier I et al (2003) Homo-FRET versus hetero-FRET to probe homodimers in living cells. Methods Enzymol 360:580–597
19. Mercier JF, Salahpour A, Angers S et al (2002) Quantitative assessment of ®1- and ®2-adrenergic receptor homo- and heterodimerization by bioluminescence resonance energy transfer. J Biol Chem 277:44925–44931
20. Zacharias DA, Violin JD, Newton AC et al (2002) Partitioning of lipid-modified monomeric GFPs into membrane microdomains of live cells. Science 296:913–916
21. Herman B, Krishnan RV, Centonze VE (2004) Microscopic analysis of fluorescence resonance energy transfer (FRET). Methods Mol Biol 261:351–370
22. Ward RJ, Pediani JD, Milligan G (2011) Ligand-induced internalization of the orexin OX(1) and cannabinoid CB(1) receptors assessed via N-terminal SNAP and CLIP-tagging. Br J Pharmacol 162:1439–1452
23. Vilardaga JP, Nikolaev VO, Lorenz K et al (2008) Direct inhibition of G protein signaling by cross-conformational switches between a2A-adrenergic and μ-opioid receptors. Nat Chem Biol 4:126–131

Chapter 12

Single Cell Analysis of Lipid Rafts

William T. Lee

Abstract

Lipid rafts are plasma membrane microdomains that serve as platforms for the assembly of proteins involved in signal transduction pathways. Given that lipid rafts are relatively resistant to cold extraction with non-ionic detergents, lipid raft associated and nonassociated proteins have been identified using biochemical methods such as sucrose-gradient density centrifugation. For identification of raft-associated proteins in individual cells, imaging methods, such as fluorescence microscopy, can be used. Detergent solubilization of non-raft regions of the plasma membrane and extraction of non-raft associated proteins are done on cells affixed to microscope slides and prior to immunostaining. This methodology has the advantages of requiring smaller cell numbers than traditional biochemical methods and also permits the study of migration of signaling proteins into and out of rafts during cell activation. An additional adaptation of the method allows identification of lipid raft-associated proteins during cognate interactions between cells. Here, as an example, we describe the methodology used in our laboratory to study lipid raft-associated molecules during T lymphocyte interactions with antigen-presenting cells.

Key words Lipid rafts, Plasma membrane, Signal transduction, Microscopy, Cell–cell interactions, T lymphocytes

1 Introduction

Lipid rafts, also called glycolipid-enriched membranes or detergent-insoluble glycosphingolipid-enriched domains, have been studied in the context of neurotransmission, regulation of membrane fluidity, and also receptor signaling and receptor trafficking in different cell types (reviewed in [1]). However, most studies have focused on their role as platforms for assembling signaling clusters in activating immune cells (B and T lymphocytes). Lipid rafts are formed through enrichment of certain membrane lipids, particularly cholesterol, glycosphingolipids, and sphingomyelin, surrounded by unsaturated glycerophospholipids [2]. These specialized plasma membrane microdomains are more ordered and tightly packed than the surrounding bilayer, but float freely in the membrane. Although most membrane proteins are absent from lipid rafts, GPI-linked and acylated proteins are concentrated in these

Troy A. Baudino (ed.), *Cell-Cell Interactions: Methods and Protocols*, Methods in Molecular Biology, vol. 1066, DOI 10.1007/978-1-62703-604-7_12, © Springer Science+Business Media New York 2013

microdomains [2]. Thus, membrane proteins may either be always excluded or constitutively associated with lipid rafts. Alternatively, proteins may migrate into rafts during cell activation. While not without controversy [3, 4], it is generally believed that such regulation of the spatial association or, alternatively, prevention of association, of interacting proteins plays a critical role in productive signal transduction. For example, in T lymphocytes engagement of the receptor for antigen (TCR) and subsequent cell activation is spatially regulated at a highly organized region (immunological synapse) formed at the interface between the T cell and the antigen-presenting cell (APC) (reviewed in [5]). Lipid rafts also migrate to the immune synapse and several lines of evidence point to them being a key regulator of signaling: (1) Engagement of the TCR by antigen or anti-TCR antibodies leads to rapid raft aggregation [6]; (2) Several of the molecules necessary for T cell signaling and activation are constitutively associated or migrate to lipid rafts upon TCR engagement [1, 6]; (3) Signaling protein activation, as measured by tyrosine phosphorylation, occurs on lipid rafts [1, 3]; (4) Raft aggregation alone, by clustering GM1 gangliosides with cholera toxin B, results in TCR-associated signaling molecule recruitment and tyrosine phosphorylation [6]; (5) Inhibition of cell activation (clonal anergy) was associated with exclusion of essential signaling molecules from lipid rafts [7, 8].

Compared to non-raft elements of the plasma membrane, lipid rafts are resistant to extraction at low temperature by nonionic detergents, such as Triton X-100 (TX-100) [9, 10]. Thus, rafts and raft-associated proteins are typically identified after separation of membrane detergent lysates by density gradient centrifugation [10, 11]. However, large numbers of cells are required for this technique and an alternative approach is necessary when starting material, such as subpopulations of primary ex vivo cells, is limiting. When material is limiting, raft-associated proteins may be identified on single cells using microscopy. We have adapted a method initially described by Janes et al. to identify lipid raft-associated, TCR-signaling proteins in resting T cells [6]. The cells of interest are affixed to microscope slides and non-raft proteins are extracted before staining the cells for target proteins of interest. Thus, by comparing cells that have been treated with detergent to cells that have not been treated, raft-associated proteins (present after treatment) and raft-excluded proteins (absent after treatment) can be identified. We found that cells of different differentiation stages (e.g., naive versus memory T cells) have different patterns of constitutive raft-associated signaling assemblies [12]. For example, Fig. 1 shows two signaling proteins (TCR and CD45) that are constitutive raft-associated proteins in memory cells (as indicated by resistance to detergent extraction) but not naive cells. This pattern might reflect the different activation capabilities of the two cell types. We have further extended this method to examine individual T cells during their cognate interactions with APCs and subsequent

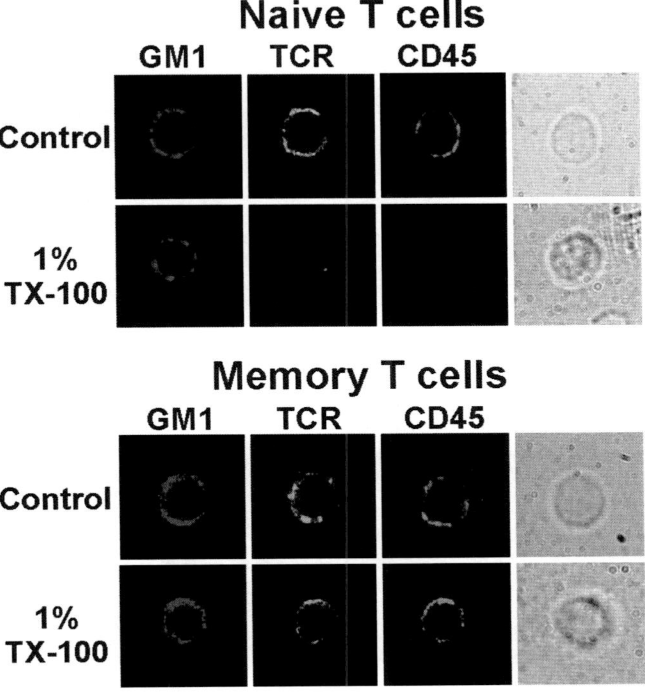

Fig. 1 Cell-type differences in constitutive raft-associated proteins. Mouse naive and memory T cells were labeled with cholera toxin-B-subunit (CTB)-rhodamine to identify lipid rafts (GM1). The cells were either (*control*) stained with fluorescent-labeled antibodies (as described in Subheading 3) or (*1 % TX-100*) permeabilized with 1 % Triton X-100 (to remove non-raft proteins) before staining. The antibodies were directed against two membrane proteins involved in cell signaling (TCR or CD45). Staining is shown with conversion to grayscale. The *far-right column* shows the differential interference contrast (DIC) images of cells

activation by specific antigen [8, 12]. We showed that regulation of lipid raft assemblies, namely exclusion of critical signaling molecules, consequently controls productive signal transduction and cell activation [8]. For example, during activation of memory T cells the critical signaling protein ZAP-70 associates with lipid rafts [1] and migrates to the immune synapse ([13] and Fig. 2). However, in memory cells exposed to a toxin that causes them to be inactivated (anergic), ZAP-70 is excluded from both lipid rafts and the immunological synapse ([8] and Fig. 2). Thus, cell activation is blocked by the physical separation of key signaling proteins.

Here we describe our laboratory's procedure for identification of lipid rafts and raft associated proteins in individual resting T cells or interacting couplets of T cells and APCs. This method relies upon the specific identification of the signaling protein of interest by staining with fluoresceinated antibodies before or after membrane extraction with cold, nonionic detergent, followed by fluorescence microscopy. Although our focus is on immune cells, the techniques are adaptable to other cell types and cell–cell interactions.

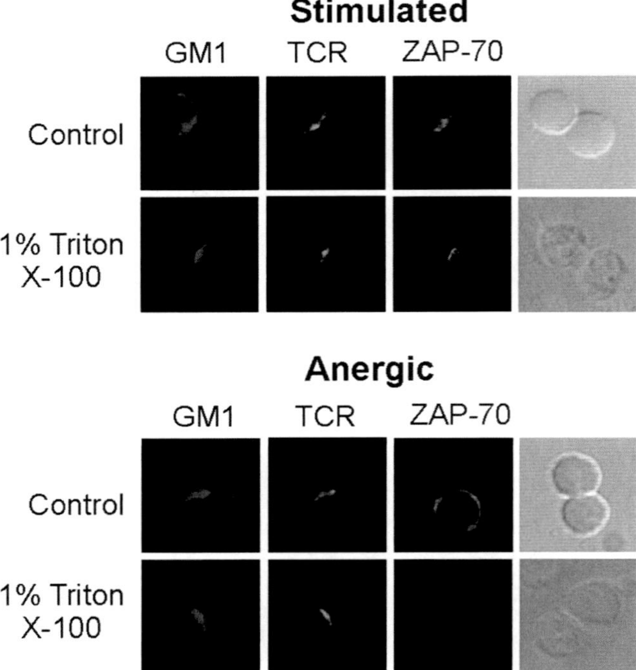

Fig. 2 Signaling blockade is associated with exclusion from lipid rafts. Memory CD4+ T cells were labeled with CTB-rhodamine (GM1) and were conjugated with antigen-presenting cells pre-pulsed with either (*Stimulated*) specific antigen, or (*Anergic*) an inactivating toxin. The cells were either (*control*) stained with fluorescent-labeled antibodies, or (*1 % TX-100*) permeabilized with 1 % Triton X-100 before staining. The antibodies were directed against two essential signaling proteins (TCR or ZAP-70). Staining is shown with conversion to grayscale. The *far-right column* shows the DIC images of the conjugates

Furthermore, the microscopy technique also allows for examination of dynamic events occurring in individual cells (or during "crosstalk" between interacting cells) as opposed to bulk populations.

2   Materials

All solutions and equipment coming into contact with cells, at least through the cell culture stage, must be sterile. Prepare all solutions using Milli-Q-purified water or equivalent. Prepare and store all reagents at 4 °C, unless otherwise indicated.

2.1   Cell Culture

1. Cells: primary ex vivo cells or tissue culture cells (*see* **Note 1**).

2. Tissue culture medium: complete RPMI-10 medium (*see* **Note 2**). RPMI-10 medium is enriched by the addition (to

final concentrations) of Fetal Bovine Serum (FBS) (10 %), L-glutamine (2 mM), 2-mercaptoethanol (50 µM), penicillin (100 U/ml), and streptomycin sulfate (100 µg/ml).

3. Stimulus to be tested.

4. 12-well tissue culture plates.

2.2 Slide Preparation

1. 12-well multitest slides.

2. 70 % ethanol.

3. 0.01 % poly-L-Lysine solution: 0.1 % poly-L-Lysine solution diluted 1:10 with deionized water (*see* **Note 3**).

4. Phosphate-Buffered Saline (PBS).

2.3 Lipid Raft Identification and Cell Staining

1. PBS/1 % bovine serum albumin (BSA) and PBS/0.1 % BSA.

2. Triton X-100 extraction buffers (2 detergent percentages): 20 mM Tris (pH 7.5), 150 mM NaCl, 1 mM EGTA, 5 mM EDTA, 10 µg/ml aprotinin, 10 µg/ml leupeptin, 10 mM iodoacetamide, 5 mM benzamidine HCl, 1 mM AEBSF, 50 mM NaF (optional, *see* **Note 4**), 10 mM $Na_4P_2O_7$ (optional), 1 mM Na_3VO_4 (optional), add Triton X-100 to either 1 % (v/v) or 0.2 % (v/v). The Triton X-100 solutions, without the protease inhibitors, can be stored for a year or longer at 4 °C. Add the protease inhibitors immediately before use from 100× stocks in water. Store the protease inhibitor stocks for up to 3–4 months at –20 °C.

3. 0.4 % paraformaldehyde solution: 4 g paraformaldehyde (electron microscopy grade) in 100 ml of PBS. Dissolve paraformaldehyde by heating solution to 70 °C in a fume hood for 1 h. Cool to room temperature and adjust pH to 7.2 with 0.1 M NaOH or 0.1 M HCl. Store protected from light for up to 2 weeks at 4 °C.

4. Pre-titrated (optimal) primary or directly conjugated fluorescent antibody (*see* **Note 5**).

5. Pre-titrated (optimal) fluorescent secondary antibody (*optional*).

6. Rhodamine (TRITC)-conjugated cholera toxin B subunit (CTB) (List Biological Laboratories, Campbell, CA, USA) or Alexa Fluor 594-conjugated CTB (Invitrogen, Grand Island, NY, USA) diluted to 10 µg/ml in PBS/0.1 % BSA (optional, *see* **Note 6**).

7. Anti-CTB antibody (EMD Millipore, Billerica, MA, USA), used at dilution suggested by manufacturer (*optional, see* **Note 6**).

8. 10 mM Methyl-β-cyclodextrin (MβCD) (Sigma-Aldrich, St. Louis, MO, USA) dissolved in PBS (*optional, see* **Note 7**).

9. SlowFade Light Antifade Kit (Invitrogen).

3 Methods

Similar to the density gradient centrifugation technique, rafts and raft-associated proteins are identified by resistance to Triton X-100 solubilization. Cells are treated with detergent after adherence to microscope slides and solubilized non-raft proteins are removed by washing. Hence, fluorescent microscopy examination of individual proteins before and after exposure to Triton X-100 indicates which proteins are associated with lipid rafts. An immunological example (T lymphocytes) is described; however, the basic procedure may be adapted to other cell types.

3.1 Slide Preparation

1. Set up four wash chambers that are large enough to hold the desired number of slides to be coated with poly-L-lysine (*see* **Note 3**). Slides should be laid flat in the chambers and can be manipulated by using forceps to grab onto one corner of the slide. The wash chambers should be filled with enough volume to fully cover the slides and the individual chambers should contain: (1) 70 % ethanol; (2) deionized water; (3) 0.01 % poly-L-lysine solution; (4) PBS.

2. Submerge the 12-well multitest slides into the chamber containing 70 % ethanol and incubate for 5 min at room temperature.

3. Remove the slides and transfer into the chamber containing deionized water and incubate for 5 min at room temperature.

4. Transfer the slides into the poly-L-lysine solution and incubate for 5 min at room temperature.

5. Remove the slides and air dry.

6. Repeat the poly-L-lysine incubation and air-drying step for a total of three incubation periods.

7. Transfer the slides to the chamber containing PBS and incubate for 5 min at room temperature.

3.2 Lipid Raft Identification in Single Cells

This procedure for examination of individual cells is useful for determining if molecules of interest are constitutively associated with lipid rafts in nonactivated cells. The procedure may also be applied to cells after cell activation or receptor–ligand interaction to determine whether the molecule of interest migrates to rafts during signaling.

1. Prepare the primary (or immune) cells of interest for study (*see* **Note 1**). Alternatively, collect cells from tissue culture (*see* **Note 8**). In these procedures we will use T cells as our model cell type.

2. (*Optional*) If lipid rafts are first labeled with fluorescent CTB for identification of the GM1 ganglioside component of rafts,

then resuspend the T cells to 2×10^7/ml in PBS/0.1 % BSA in a 15 ml centrifuge tube (*see* **Notes 6** and **9**). Then add CTB-rhodamine or CTB-Alexa Fluor 594 to 10 µg/ml. Incubate the tube for 30 min on ice. Fill the centrifuge tube PBS/0.1 % BSA and centrifuge at $200 \times g$ for 10 min. Repeat the wash step. Wash an additional three times with 15 ml of PBS (no BSA). If CTB labeling is not desired, then omit this step and move to **step 3**.

3. Prepare the slides by washing the cells, if they were not washed in **step 2**, with 15 ml of PBS (no BSA). Resuspend the T cells to 2×10^7/ml in PBS and then pipet 20 µl of cells into each well of the poly-L-lysine-coated 12-well multitest slides and then incubate for 20 min in a 37 °C humidified incubator (*see* **Note 10**). Use forceps to pick the slides up at one corner and, using a squirt bottle containing a washing buffer (e.g., PBS), project a steady stream onto the slide surface for a minimum of 10 s. Shake the slide free of wash buffer. This constitutes a single wash, which should be repeated for a total of three washes.

4. (*Optional*) If patching or aggregation of lipid rafts is desired (*see* **Note 6**), then add this step into your protocol. If not desired, then proceed to **step 5**. To patch lipid rafts, wash the slides once with PBS/0.1 % BSA, as described in **step 3**, and then add 15 ml of anti-CTB antibody (1:250 dilution of the commercial preparation in PBS/0.1 % BSA) to the sample well(s). Incubate the slides for 30 min on ice. Raise the temperature of the cells by moving the slides to 37 °C and continue incubating for an additional 20 min.

5. Extract the non-raft proteins by selecting wells designated as the control samples and the wells designated as the test samples (*see* **Note 11**). For control samples, fix the cells by adding 15 µl of 4 % paraformaldehyde to each well and then incubating the slides for 20 min at room temperature. Wash the slides once with PBS as described in **step 3**, and then permeabilize the cells by adding 15 µl/well of the 0.2 % TX-100 solution. Incubate the slides for 5 min at 4 °C and then wash the slides once with PBS as described in **step 3**. (*Optional*) If MβCD treatment is to be done (*see* **Note 7**), *prior* to the cell fixation step, incubate the slides with a 10 mM MβCD solution for 30 min at 37 °C and wash the slides twice with PBS as described in **step 3**.

6. For test samples, permeabilize the cells by adding 15 µl/well of the 1 % TX-100 solution and incubating the slide for 5 min at 4 °C. Wash the slides once with PBS as described in **step 3**. Fix the cells by adding 15 µl of 4 % paraformaldehyde to each well and then incubating the slides for 20 min at room temperature. Wash the slides once with PBS as described in **step 3**. (*Optional*) If MβCD treatment is to be done (*see* **Note 7**), *prior* to permeabilization with the 1 % TX-100 solution, incubate the

slides with a 10 mM MβCD solution for 30 min at 37 °C and wash the slides twice with PBS as described in **step 3**.

7. To identify proteins by immunostaining, first block the slides by incubating them PBS/1 % BSA overnight at 4 °C (*see* **Note 12**). The next day, bring the slides to room temperature and wash them twice with PBS as described in **step 3**. Stain the cells with 15 μl of either the unlabeled primary antibody or the directly conjugated fluorescent antibody at the optimal dilution for 1 h at room temperature (*see* **Note 13**). Wash slides three times with PBS as described in **step 3**. If you are using indirect staining, then stain the cells with 15 μl of the fluorescent secondary antibody at its optimal dilution for 1 h at room temperature. Wash the slides three times with PBS as described in **step 3** and then mount the coverslips using the SlowFade Antifade Kit according to the manufacturer's instructions. Store the slides in the dark at 4 °C until microscopy analysis (*see* **Note 14**).

8. Analyze cells using fluorescent-based microscopy (*see* **Note 15**). Compare extraction control and test samples for each cell treatment group.

3.3 Lipid Raft Identification During Cell–Cell Interactions

T cell stimulation by peptide antigen requires that the T cell physically interact with another cell (APC) for proper activation. This procedure allows for the study of movement of lipid rafts and the migration of raft-associated proteins during cell–cell interactions. The steps followed in this method are similar to those used in Subheading 3.2 above, except that prior to processing for microscopy, cell–cell conjugates are formed. APCs bind to specific antigen and then couple with specific T cells to induce T cell activation. Conjugates containing both cell types are formed and then adhered to microscope slides and cultured to permit cell signaling.

1. Prepare cells of interest for study (*see* **Note 1**). In this immunological example, T cells must recognize the specific antigen that will be presented and APCs must be MHC histocompatible (*see* **Note 16**). T cells and APCs are prepared separately. Primary CD4+ T cells or T cell subsets are prepared. Alternatively, collect and process T cells from tissue culture. Primary APCs are prepared or collected from tissue culture. Either cell type may be held in complete RPMI-10 medium on ice in 15 ml conical tubes until used.

2. (*Optional*) If lipid rafts are first labeled with fluorescent CTB for identification of the GM1 ganglioside component of rafts, then only label the cell type of interest (e.g. T cells), while leaving the partner cell population unlabeled. Thus, resuspend the T cells to 2×10^7/ml in PBS/0.1 % BSA in a 15 ml centrifuge tube (*see* **Notes 6** and **9**). Then add CTB-rhodamine or

CTB-Alexa Fluor 594 to a concentration of 10 µg/ml. Incubate the tube for 30 min on ice. Fill the centrifuge tube with PBS/0.1 % BSA and centrifuge at $200 \times g$ for 10 min. Repeat the wash step and then wash three more times with 15 ml of PBS (no BSA). If CTB labeling is not desired, omit this step and move to **step 3**.

3. To pulse the APCs with antigen, resuspend the cells to $1 \times 10^7/$ml in prewarmed 37 °C complete RPMI-10 medium with 1 µg/ml of peptide antigen and add the cells into individual wells of a 24-well tissue culture plate (1 ml/well) (*see* **Note 17**). Incubate the culture plate for 2 h at 37 °C in a 5 % CO_2 tissue culture incubator (*see* **Note 10**). After the incubation period, collect the cells and transfer them into a 15 ml tube. Centrifuge for 10 min at $200 \times g$ at 4 °C. Resuspend the cells in 15 ml PBS and centrifuge again for 10 min at $200 \times g$, 4 °C. Resuspend the APCs to $4 \times 10^7/$ml in PBS.

4. Resuspend the T cells to $2 \times 10^7/$ml in prewarmed (37 °C) PBS.

5. To form conjugates between the antigen-specific T cells and the antigen-pulsed (or control) APCs, add 12.5 µl antigen-pulsed APCs and 12.5 µl of T cells to 1.5 ml microcentrifuge tubes and gently pipet up and down several times to mix the cells. Centrifuge the tubes at $400 \times g$ for 10 s at room temperature. Quickly and carefully pipet 10 µl of cells onto prewarmed (37 °C) poly-L-lysine coated wells of a 12-well multitest slide and incubate at 37 °C for various lengths of time (0–60 min) (*see* **Note 18**). Wash the slides once with PBS as described in Subheading 3.2, **step 3**.

6. Identify samples not only as stimulation test samples and controls but also as paired samples for non-raft protein extraction test samples and control samples as described in Subheading 3.2, **step 5** (*see* **Note 11**). For extraction control samples, fix the cell conjugates by adding 15 µl of 4 % paraformaldehyde to each well and then incubating the slides for 20 min at room temperature. Wash the slides once with PBS as described in Subheading 3.2, **step 3**, then permeabilize the cells by adding 15 µl/well of the 0.2 % TX-100 solution. Incubate the slides for 5 min at 4 °C and then wash the slides once with PBS as described in Subheading 3.2, **step 3** (*see* **Note 19**).

7. For test samples, first permeabilize the cell conjugates by adding 15 µl/well of the 1 % TX-100 solution and incubating the slide for 5 min at 4 °C. Wash the slides once with PBS as described in Subheading 3.2, **step 3**, then fix the cells by adding 15 µl of 4 % paraformaldehyde to each well and incubating the slides for 20 min at room temperature. Wash the slides once with PBS as described in Subheading 3.2, **step 3**.

8. To identify proteins in the conjugates by immunostaining, first block the slides by incubating them PBS/1 % BSA overnight at 4 °C (*see* **Note 12**). The next day, bring the slides to room temperature and wash them twice with PBS as described in Subheading 3.2, **step 3**. Stain the cells with 15 μl of either the unlabeled primary antibody or the directly conjugated fluorescent antibody at the optimal dilution for 1 h at room temperature (*see* **Note 13**). Wash the slides three times with PBS as described in Subheading 3.2, **step 3**. If using indirect staining, then stain the cells with 15 μl of the fluorescent secondary (detecting) antibody at its optimal dilution for 1 h at room temperature. Wash the slides three times with PBS as described in Subheading 3.2, **step 3** and then mount the coverslips using the SlowFade Antifade Kit according to the manufacturer's instructions. Store the slides in the dark at 4 °C until microscopy analysis (*see* **Note 14**).

9. Analyze by fluorescence-based microscopy (*see* **Note 15**). Compare extraction control and test samples for each cell treatment group.

4 Notes

1. Obviously this will vary depending on the individual laboratory's specific experimental model. In our laboratory, we generally begin with single cell suspensions from murine spleen or lymph nodes after removal of red blood cells. Most often, because our own studies focus on antigen-driven T lymphocyte responses, we enrich our starting populations for either CD45RB^{lo} (antigen-experienced or memory) and CD45RB^{hi} (naive) CD4 T cells, using methods that we previously have described [12, 14]. Very few cells are needed for an individual test.

2. The percentage of serum used in Subheading 3 is indicated by medium-numeral. In this example, RPMI-10 indicates that FBS is added to 10 %. If no numeral is added to the base name, no serum is added. The FBS is heat inactivated for 1 h at 56 °C and stored at 4 °C before use. The base medium, RPMI, is routinely used in our laboratory for tissue culture; we have not tested other media types but it is likely that the routine culture media for the user is suitable for this procedure.

3. Coating the 12-well multitest slides with poly-L-lysine will increase cell attachment to the slide. The volume of poly-L-lysine solution will vary depending upon the number of slides to be coated. The slides will be immersed in the solution, so sufficient coating solution should be prepared to cover all slides to be used in the experiment.

4. Phosphatase inhibitors (NaF, $Na_4P_2O_7$, Na_3VO_4) should be included when examining phosphoproteins or raft associations that occur as a consequence of kinase activity.

5. We routinely use Alexa dyes because of their brighter and more stable fluorescence as compared to other dyes. However, we have had good success using more commonly used dyes such as FITC, Rhodamine, and Cy5. Specific dilutions of the antibody reagent will vary depending upon the antigen target, manufacturer, etc., so precise concentrations are not listed here. However, all of the antibody reagents, and especially, due to their high fluorescence intensity, the Alexa-conjugated reagents, should be pre-titrated prior to assay for use at optimal concentrations.

6. Cholera toxin B subunit binds to GM1 gangliosides, which are integral components of lipid rafts [15], and therefore, using fluorescent CTB will permit visual identification by fluorescent microscopy. Generally the labeling will show a homogeneous distribution pattern because of the small size of lipid rafts (<70 nm in diameter [16]). Aggregating rafts with anti-CTB causes a patched distribution and often a better visualization of colocalization of raft-associated proteins can be seen [6]. Of note, Janes et al. demonstrated that (warm) raft aggregation by anti-CTB antibodies can promote receptor signaling [1]. We routinely include fluoresceinated CTB either along with antibodies to our protein of interest (e.g. Fig. 1) or in separate samples as part of our control group.

7. Lipid raft-associated proteins, unlike cytoskeletal or detergent-insoluble proteins, become detergent soluble in the absence of lipid rafts [17]. MβCD disrupts lipid rafts by depleting membrane cholesterol and this permits solubilization of raft-associated proteins by TX-100. We often include this treatment of cells as an additional control when determining if a specific protein is a raft protein as opposed to a protein attached to the cytoskeleton because in the former case, the protein will no longer be visualized after TX-100 treatment if the cells are first exposed to MβCD. Figure 3 shows staining of GM1 gangliosides with CTB-rhodamine with and without MβCD prior to TX-100 extraction.

8. Cell lines carried in vitro are not generally resting cells. Thus, a given experiment may be influenced by variances in culture conditions and cell growth. For T cell lines, we generally collect cells >7 days following the last stimulation period with antigen so that the cells might more reasonably reflect a resting or unstimulated cell.

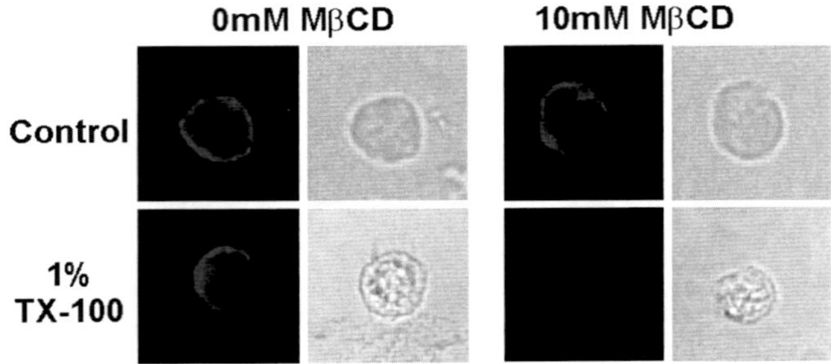

Fig. 3 MβCD disrupts lipid rafts. T cells were labeled with CTB-rhodamine to identify lipid rafts (GM1). The cells were left untreated or treated with 10 mM MβCD for 30 min at 37 °C before analysis. Staining is shown with conversion to grayscale. The *second* and *fourth columns* show the DIC images of the cells

9. The volumes and size containers described in this procedure will vary depending on the individual experiment. For our studies, we generally obtain $2.5–3.5 \times 10^7$ CD4+ T cells per one mouse spleen and may only obtain 5×10^6 CD4+ memory cells from several spleens. For our purposes, the volumes and container sizes noted in Subheading 3 are generally sufficient for our needs, including the higher volumes used for washing steps. However, these can both be scaled up depending upon the requirements of the specific experiment.

10. For all 37 °C incubations, we use our standard 5 % CO_2 culture incubator. The CO_2 is not a necessary component to the staining incubations; we use the incubator out of convenience. For any ligand stimulation or activation of the cells prior to preparation of the cells for lipid raft analysis, the tissue culture incubator is an essential component.

11. Lipid raft-associated proteins will be indicated by resistance to extraction with 1 % TX-100. In our studies, wells containing identical samples are divided into two groups to compare staining of cells with and without extraction. In Subheading 3, "control samples" refer to the staining and identification of the protein of interest in samples without extraction, whereas, "test samples" refer to the staining and identification of the same protein after extraction. For control samples, a low concentration (0.2 %) of TX-100 is added prior to the fixation step to allow access of the staining antibodies to intracellular proteins. This concentration of detergent does not cause extraction of non-raft proteins. It is essential that all reagents be kept at cold temperatures (<4 °C) and membrane solubilization must be done on ice. At higher temperatures, raft proteins become increasingly soluble [18].

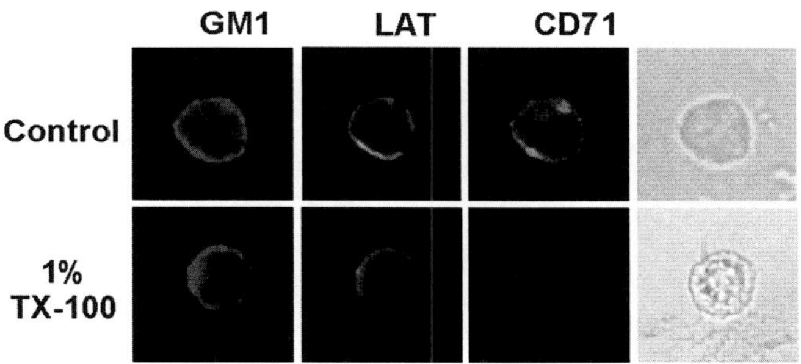

Fig. 4 Lipid raft controls. T cells were labeled with CTB-rhodamine to identify lipid rafts (GM1). The cells were either (*control*) stained with fluorescent-labeled antibodies or (*1 % TX-100*) permeabilized with 1 % Triton X-100 (to remove non-raft proteins) before staining. The antibodies were directed against a constitutively raft-associated protein (LAT) or a constitutively raft-excluded protein (CD71). Staining is shown with conversion to grayscale. The *far-right column* shows the differential interference contrast (DIC) images of cells

12. Although we usually perform an overnight blocking step in order to minimize nonspecific binding before immunostaining, it may be possible to reduce this blocking period by several hours by performing this step at room temperature.

13. Control samples should be carefully chosen and included. Important controls would be the inclusion of lipid raft measurements (e.g., CTB) and also known raft-associated proteins and known non-raft proteins. For T cell analyses, we and others have found that the proteins LAT and CD71 (transferrin receptor) work well as indicators of raft and non-raft proteins, respectively [6, 10, 12]. Figure 4 shows an example of immunostaining for LAT and CD71 with and without extraction with 1 % TX-100.

14. In our experience, fluorescence stability is enhanced for any of the labels by the use of a commercial Slow-Fade reagent. We have also found that the fluorescence integrity of the prepared slides is relatively stable when they are stored in the dark at 4 °C. However, for raft analyses the prior detergent extraction may lead to some sample instability and it is recommended that the samples be analyzed within a few days after preparation.

15. Analysis requires access to confocal or epifluorescence microscopes and deconvolution software. In our experiments, we generally acquire images using a Hamamatsu ORCA-ER digital CCD camera (Hamamatsu Photonics, Hamamatsu City, Japan) attached to a Zeiss Axioskop2 mot plus microscope (Carl Zeiss, Gottingen, Germany), using OpenLab software (Improvision Inc, Lexington, MA, USA).

16. For primary cells, wild-type, conventional mice possess too low a frequency of antigen-specific cells to be useful for this procedure. For this reason, T cells which are derived from mice transgenic for a specific TCR are most amenable for study. However, mice transgenic for a given TCRβ chain or even nontransgenic mice may be used if the presenting antigen is a bacterial superantigen [19]. Owing to the high frequency of superantigen-reactive T cells [19], cell conjugates should be visualized. Cloned T cells or antigen-specific T cell lines can also be easily used in this assay. For the analysis of CD4+ T cell-APC conjugates, APCs are prepared by treatment of murine red blood cell-depleted spleen cells with anti-Thy-1 plus baby rabbit complement to deplete T cells and enrich for B cells and macrophages [14].

17. The exact dose of antigen or other stimulus will vary for the specific antigen or stimulus. The concentration used in this procedure is for stimulation with a specific peptide antigen and was chosen so that it would lead to a strong binding interaction between the T cell and APC. A whole antigen which requires processing into peptides might require ten or even a hundred-fold higher amount to facilitate cell coupling. In our studies, we include control APCs that are either pulsed with an irrelevant (to the specific T cell of interest) peptide or unpulsed.

18. Because the T cell–APC interactions and cell signaling are dynamic events, kinetic experiments are generally done where the length of time that the T cell and APC are cultured together is varied (from seconds to hours). During this period, many molecules on both the T cell and the corresponding APC will migrate into defined membrane regions (immunological synapse) to interact in a highly organized fashion with other molecules on the same cell or on the opposing cell. Lipid rafts, visualized by CTB binding to GM1 gangliosides, also migrate into the immunological synapse [12, 20]. In this way, signaling complexes are built and signal transduction occurs in both the T cell and APC.

19. If desired, the cells can be treated with MβCD prior to extraction so that lipid rafts are disrupted and raft-association of the protein of interest is confirmed. The procedure steps would be done as described in Subheading 3.2, **steps 5** and **6** (also *see* **Note 7**).

Acknowledgment

This work was supported by National Institutes of Health (AI35583).

References

1. Janes PW, Ley SC, Magee AI et al (2000) The role of lipid rafts in T cell antigen receptor (TCR) signalling. Semin Immunol 12:23–34

2. Dykstra M, Cherukuri A, Sohn HW et al (2003) Location is everything: lipid rafts and immune cell signaling. Annu Rev Immunol 21:457–481

3. Pizzo P, Giurisato E, Tassi M et al (2002) Lipid rafts and T cell receptor signaling: a critical re-evaluation. Eur J Immunol 32:3082–3091

4. Munro S (2003) Lipid rafts: elusive or illusive? Cell 115:377–388

5. Bromley SK, Burack WR, Johnson KG et al (2001) The immunological synapse. Annu Rev Immunol 19:375–396

6. Janes PW, Ley SC, Magee AI (1999) Aggregation of lipid rafts accompanies signaling via the T cell antigen receptor. J Cell Biol 147:447–461

7. Hundt M, Tabata H, Jeon MS et al (2006) Impaired activation and localization of LAT in anergic T cells as a consequence of a selective palmitoylation defect. Immunity 24:513–522

8. Lee WT, Prasad A, Watson AR (2012) Anergy in CD4 memory T lymphocytes. II. Abrogation of TCR-induced formation of membrane signaling complexes. Cell Immunol 276:26–34

9. Brown DA, London E (2000) Structure and function of sphingolipid- and cholesterol-rich membrane rafts. J Biol Chem 275:17221–17224

10. Brown DA, Rose JK (1992) Sorting of GPI-anchored proteins to glycolipid-enriched membrane subdomains during transport to the apical cell surface. Cell 68:533–544

11. Parton RG, Simons K (1995) Digging into caveolae. Science 269:1398–1399

12. Watson AR, Lee WT (2004) Differences in signaling molecule organization between naive and memory CD4+ T lymphocytes. J Immunol 173:33–41

13. Lee KH, Holdorf AD, Dustin ML et al (2002) T cell receptor signaling precedes immunological synapse formation. Science 295:1539–1542

14. Lee WT, Cole-Calkins J, Street NE (1996) Memory T cell development in the absence of specific antigen priming. J Immunol 157: 5300–5307

15. Fra AM, Williamson E, Simons K et al (1994) Detergent-insoluble glycolipid microdomains in lymphocytes in the absence of caveolae. J Biol Chem 269:30745–30748

16. Varma R, Mayor S (1998) GPI-anchored proteins are organized in submicron domains at the cell surface. Nature 394:798–801

17. Scheiffele P, Roth MG, Simons K (1997) Interaction of influenza virus haemagglutinin with sphingolipid-cholesterol membrane domains via its transmembrane domain. EMBO J 16:5501–5508

18. Melkonian KA, Ostermeyer AG, Chen JZ et al (1999) Role of lipid modifications in targeting proteins to detergent-resistant membrane rafts. Many raft proteins are acylated, while few are prenylated. J Biol Chem 274:3910–3917

19. Janeway CA Jr, Yagi J, Conrad PJ et al (1989) T-cell responses to Mls and to bacterial proteins that mimic its behavior. Immunol Rev 107:61–88

20. Burack WR, Lee KH, Holdorf AD et al (2002) Cutting edge: quantitative imaging of raft accumulation in the immunological synapse. J Immunol 169:2837–2841

Chapter 13

Micropatterning Cell Adhesion on Polyacrylamide Hydrogels

Jian Zhang, Wei-hui Guo, Andrew Rape, and Yu-li Wang

Abstract

Cell shape and substrate rigidity play critical roles in regulating cell behaviors and fate. Controlling cell shape on elastic adhesive materials holds great promise for creating a physiologically relevant culture environment for basic and translational research and clinical applications. However, it has been technically challenging to create high-quality adhesive patterns on compliant substrates. We have developed an efficient and economical method to create precise micron-scaled adhesive patterns on the surface of a hydrogel (Rape et al., Biomaterials 32:2043–2051, 2011). This method will facilitate the research on traction force generation, cellular mechanotransduction, and tissue engineering, where precise controls of both materials rigidity and adhesive patterns are important.

Key words Cell adhesion, Cell geometry, Substrate rigidity, Mechanotransduction

1 Introduction

Cell–cell interactions involve both the release of chemicals and the generation of mechanical forces; the latter includes forces transmitted through both cell–cell junctions [1] and cellular adhesions to the elastic extracellular matrix (ECM). Mechanical forces applied by contractile cells adhered to the ECM, termed traction forces, may propagate much more rapidly and efficiently than chemical signals to affect distant cells that adhere to the same matrix [2]. These mechanical signals are known to regulate a range of cellular activities including migration, proliferation, apoptosis, and differentiation [3–10]. In addition, adherent cells also use traction forces to probe physical properties of the environment such as rigidity and topography [11], such that long-lasting cell–cell interactions may be mediated by modifying physical characteristics of the ECM.

Due to their optical transparency, cost-effectiveness, ease of preparation, and controllable elasticity over the physiological range, polyacrylamide (PAA) gels have become widely used as a tool for measuring cellular traction forces and probing cellular responses to

Troy A. Baudino (ed.), *Cell-Cell Interactions: Methods and Protocols*, Methods in Molecular Biology, vol. 1066, DOI 10.1007/978-1-62703-604-7_13, © Springer Science+Business Media New York 2013

mechanical cues [12]. The lack of passive protein adsorption on PAA further allows the control of cell shape, spreading size, and migration by conjugating defined surface areas with specific ECM ligands. In principle, patterning of PAA may be achieved by micro-contact printing ECM proteins onto a gel surface that has been chemically activated with agents such as the hetero-bifunctional crosslinker sulfo-SANPAH [12]. However, the deformability of both the stamp and the gel surface makes this approach poorly reproducible when applied to PAA [13, 14] or polydimethylsiloxane (PDMS; [15]). Alternatively, the surface of a hydrogel may be micropatterned at a high resolution using a combination of photochemistry and confocal laser scanning optics [16]; however, this approach is inefficient for patterning large surface areas.

We have developed a simple and efficient method for conjugating proteins to the surface of PAA hydrogels at a high resolution [17]. The method takes advantage of the extensive glycosylation of many ECM proteins, which contain vicinal diols that may be chemically activated with sodium m-periodate to form two aldehyde groups after a ring-opening reaction [13, 18]. The aldhehyde groups in turn react with polymerizing acrylamide to incorporate the ECM proteins into the PAA gel structure. In addition to the compatibility with a micropatterning procedure as described below, this simple approach is easily repeatable and requires much less time than previously reported methods (30 min compared to up to 4 h).

There are many potential applications of micropatterned hydrogels. Constraining cell shape proved useful in traction force measurements. Since traction forces vary with cell shape [15, 19, 20], eliminating the variability of cell shape reduces the standard deviation and allows the detection of previously obscure differences. More importantly, this method enables researchers to assess relative influences of cell shape and substrate rigidity on such processes as differentiation, apoptosis, and proliferation. For the study of cell migration, this method may also be used to constrain cells to take defined shapes and paths. Combinatorial manipulations of multiple parameters on micropatterned PAA will likely lead to new advances in both cell biology and tissue engineering.

2 Materials

2.1 Activation of Coverslips

1. Freshly prepared Bind-silane working solution. Mix 950 µl of ethanol (95 %; ACS/USP grade) and 50 µl of acetic acid with 3 µl of Bind-silane (Sigma-Aldrich, St. Louis, MO, USA).

2. Coverslips (45 mm × 50 mm, No. 1).

3. Diamond-tip pen.

4. Bunsen burner.

2.2 Photolithography	1. SPR-220.3 positive photoresist (MicroChem, Newton, MA, USA).

1. SPR-220.3 positive photoresist (MicroChem, Newton, MA, USA).

2. Coverslips (45 mm × 50 mm, No. 2).

3. Spin coater (e.g., Spin Processor from Laurell, North Wales, PA, USA). An inexpensive low-speed tabletop centrifuge may be modified to hold coverslips for spin coating [21].

4. Heating block or plate with precise temperature control for 115 °C.

5. UV source for i-line (365 nm). An inexpensive UV station with relatively uniform exposure may be constructed by mounting a high flux UV LED (Opto Technology Inc, Wheeling, IL, USA) over an orbit shaker where the photoresist-coated coverslip is placed. The setup is shown in [21].

6. Photomask with the desired pattern. Glass photomasks (e.g., from Advance Reproductions Corp, North Andover, MA, USA) are required for a resolution of pattern better than 10 μm. Otherwise, the mask may be ordered as an inexpensive transparency film (e.g., from CAD/Art Services, Bandon, OR, USA).

7. Microposit Developer MF-319 (MicroChem).

8. Glass Petri dishes.

9. Orbital shaker in a chemical fume hood.

2.3 Preparation of PDMS Stamps

1. Sylgard 184 Silicone Elastomer Kit, including Base and Curing Agent (Dow Corning Corporation, Midland, MI, USA).

2. Balance and weighing boats.

3. Heating block or incubator with precise temperature control for 70 °C.

4. Vacuum.

5. Disposable beakers.

2.4 Activation of ECM Proteins

1. Gelatin or ECM protein, such as collagen or fibronectin.

2. Sodium m-periodate (Sigma-Aldrich).

2.5 Micropatterning of Polyacrylamide Hydrogels

1. Glass coverslips (25 mm × 25 mm, No. 1).

2. Acrylamide.

3. Bisacrylamide.

4. HEPES: 200 mM, pH 8.5.

5. Nitrogen gas.

6. Ammonium persulfate (APS), freshly prepared 10 % (w/v) solution.

7. N,N,N',N'-Tetramethylethylenediamine (TEMED).

8. Razor blades.

9. (For traction force microscopy): Fluorescent latex beads of different colors, such as Fluoresbrite carboxy microspheres, 0.2 μm, Yellow/Green (Polysciences, Warrington, PA, USA), and FluoSpheres carboxylate-modified microspheres: 0.1 μm, blue fluorescence (350/440) (Molecular Probes, Eugene, OR, USA).

3 Methods

3.1 Activation of Coverslips for the Bonding of PAA

1. Mark one side of the 45 mm × 50 mm, No.1 coverslip with a diamond-tip pen. Pass the coverslip over the inner flame of a Bunsen burner with the marked side facing the flame. The plasma in the flame increases the hydrophilicity of the glass surface. Allow the coverslips to cool to room temperature.

2. In a fume hood, apply approximately 30 μl of Bind-silane working solution onto the flamed side of coverslips and smear it evenly with the pipet tip. Remove excess Bind-silane with Kimwipes. Allow the Bind-silane to react for 3 min.

3. Rinse treated coverslip surfaces with ethanol and wipe with Kimwipes to remove any residual Bind-silane solution. Allow to air-dry (*see* **Note 1**).

3.2 Preparation of PDMS Stamps

1. Pass coverslip (45 mm × 50 mm, No. 2) over the inner flame of a Bunsen burner with the marked side facing the flame. Allow the coverslip to cool to room temperature.

2. In a fume hood with a spin coater set at 5,000 rpm for 30 s, spread 180 μl positive photoresist SPR-220 uniformly across the flamed side of the coverslip.

3. Bake the coverslips at 115 °C for 90 s on a heating block. Allow the coverslips to cool to room temperature.

4. Place the photomask over the coverslip and expose the assembly to 365 nm UV light. The exposure depends on the intensity of the light source. Using a high flux UV LED at a distance of 3 cm from the coverslip on an orbital shaker, the optimal exposure is around 45 s (*see* **Note 2**).

5. Bake the coverslip at 115 °C for 90 s on a heating block. Allow the coverslip to cool to room temperature.

6. Immerse the coverslip in Microposit Developer MF-319 in a glass Petri dish, placed on an orbital shaker inside a chemical fume hood, for approximately 45 s. Optimal timing and mixing conditions are affected by the exposure condition and should be controlled carefully (*see* **Note 3**).

7. Rinse the coverslip extensively in deionized water and allow to air-dry. The pattern on SPR-220, which serves as the molding for PDMS stamps, should be visible (*see* **Note 4**).

8. Weigh out approximately 5 g of Sylgard 184 Silicone Elastomer Base, add 1/10 volume (v/w) of the curing agent, and mix thoroughly. Degas for 30 min using house vacuum to remove air bubbles.

9. Incubate the coverslip covered with Sylgard at 70 °C on a heating block or in an incubator for at least 1 h.

3.3 Activation of ECM Proteins

1. Dilute gelatin to a final concentration of 0.1 % (w/v) in PBS, or Type I collagen to a final concentration of 0.01 % (w/v) in 50 mM sodium acetate (pH 4.5), or fibronectin to a final concentration of 0.001 % (w/v) in PBS.

2. While mixing, slowly add solid sodium *m*-periodate to the protein solution to reach a final concentration of 3.6 mg/ml. Allow the reaction to proceed for 30 min at room temperature (*see* **Note 5**).

3.4 Micropatterning of Polyacrylamide Gels

1. Prepare the PDMS stamp, activated protein, and Bind-silane-activated glass coverslips as described above (Subheadings 3.1, 3.2, and 3.3).

2. Pass a 25 mm × 25 mm coverslip over the inner flame of a Bunsen burner. Allow the coverslip to cool to room temperature.

3. Pipet approximately 200 µl of the activated protein solution onto the surface of the stamp, incubate at room temperature for 30 min, then remove excess solution by blowing with a stream of nitrogen gas.

4. Prepare the acrylamide solution with the desired concentrations of acrylamide and bisacrylamide in 10 mM HEPES. Typically 5 % (w/v) acrylamide and 0.1 % (w/v) bisacrylamide are used for the measurement of traction stress of fibroblasts. Degas with house vacuum for 20 min, as previously described [12].

5. Press the stamp against the 25 mm × 25 mm coverslip for 5 min (*see* **Note 6**).

6. Add 0.006 volumes of 10 % APS and 0.004 volumes of TEMED to the acrylamide solution and mix quickly and briefly to initiate the polymerization reaction.

7. Pipet 30 µl of the polymerizing acrylamide solution onto a Bind-silane-activated coverslip.

8. Without any delay, remove the stamp from the 25 mm × 25 mm coverslip and place the coverslip, stamped side facing down, on the polymerizing acrylamide solution.

9. Let the acrylamide polymerize to completion for 15–20 min at room temperature.

10. Peel off the top coverslip carefully with a razor blade. Cover the hydrogel surface immediately with PBS to prevent drying (*see* **Notes** **7** and **8**).

3.5 Preparation of Micropatterned Polyacrylamide Substrates for Traction Force Microscopy

1. Prepare the coverslip stamped with activated gelatin (or other ECM proteins) as described above (Subheading 3.4).

2. Prepare and degas the acrylamide solution as described above (Subheading 3.4).

3. Add 0.006 volume of 0.1 % (w/v) gelatin to the acrylamide solution. This is required for obtaining a uniform distribution of beads (*see* **Note** **9**).

4. Add a dilution of 0.0008 volume of 0.2 μm (*see* **Note** **10**) fluorescent latex beads into the acrylamide solution. Mix well and incubate at room temperature for 2.5 min (*see* **Note** **11**).

5. Add 0.006 volume of 10 % (w/v) APS, 0.004 volume of TEMED, and 0.002 volume of 0.1 μm fluorescent beads of a different color from that used in **Step 4** (optional, *see* **Note** **12**). Mix rapidly and pipet 30 μl on the surface of Bind-silane-activated coverslip.

6. Immediately place the stamped coverslip, patterned side down, on the acrylamide solution (*see* **Note** **13**). Let the acrylamide polymerize to completion for 15–20 min at room temperature. Peel off the top coverslip carefully with a razor blade. Cover the hydrogel surface immediately with PBS to prevent drying (*see* **Note** **14**).

4   Notes

1. Activated coverslips may be stored in a desiccator at room temperature for at least 3 months.

2. Tight contact between the photomask and photoresist is critical if the light source is not well collimated. A simple method involves placing the photomask over the coverslip and sandwiching them between two pieces of glass plates with paper clamps to ensure tight contact. In addition, it is critical to make sure that the patterned side on the photomask is facing the coverslip. For illumination with a less-than-uniform light source, the assembly of photomask and coverslip may be

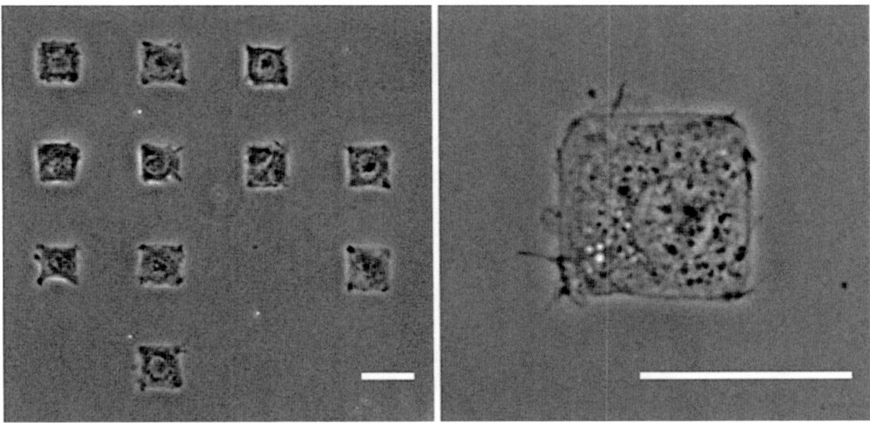

Fig. 1 MC3T3-E1 cells on PAA hydrogel patterned as a *square* with gelatin. Scale bar, 50 μm

placed on an orbit shaker [21]. Rotation at about 70 rpm during exposure creates a uniform average illumination across the surface. The exact exposure time depends on the optical condition and must be calibrated for each setup.

3. Developing in tetramethylammonium hydroxide (TMAH; 2.45 % in 0.1 % Triton X-100) generates results comparable to those using Microposit Developer MF-319.

4. To generate polyacrylamide substrates uniformly conjugated with ECM, simply use a clean coverslip to make a flat, patternless PDMS stamp.

5. Treatment with periodate causes vicinal diols in the sugar moieties of ECM proteins to undergo a ring-opening reaction, forming two aldehyde groups [13, 18], which are capable of copolymerizing with acrylamide, thus directly incorporating the ECM protein into the hydrogel. After the reaction, periodate can be removed with dialysis or a spin column. Activated proteins may be stored at –20 °C or –80 °C for several months.

6. The contact between the stamp and the coverslip should be even to ensure efficient transfer of the patterned protein. Even contact may be achieved by gently rolling a pencil back and forth over the PDMS stamp in two criss-cross directions. The quality of contact may be checked by looking at the glass–PDMS interface from different angles. Colorful interference fringes appear at the interface if the contact is good.

7. ECM-micropatterned hydrogel substrate should be sterilized using UV irradiation before inoculation with cells (Fig. 1).

8. Non-glycosylated proteins may be patterned using a similar procedure. The protein is stamped onto the coverslip as described above without activation. Acrylamide solution is prepared with the incorporation of acrylic acid-*N*-hydroxysuccinimide, which reacts with protein lysine residues, at a final concentration of 0.05 mg/ml.

9. We have found that fluorescent latex beads became concentrated onto protein conjugated areas during acrylamide polymerization, allowing easy visualization of the pattern but creating problems for the measurement of substrate strain for traction force microscopy. Addition of gelatin to the acrylamide solution prevents this concentration, suggesting electrostatic interactions as the cause of bead concentration. Proteins other than gelatin, e.g., BSA, may also be used to block beads concentration.

10. This beads concentration worked well to generate an optimal surface bead density for traction force microscopy (i.e., high enough to provide a good resolution while allowing the resolution of individual beads). The concentration may be adjusted according to the requirement of the experiment.

11. The incubation allows the protein (gelatin) in the solution to bind to the beads. Too short of an incubation provides insufficient blocking and causes beads to accumulate under protein conjugated areas. Conversely, too long an incubation causes beads to be excluded from protein conjugated areas, likely due to the repulsive electrostatic interactions. Blocking time may be influenced by protein and beads concentration, and needs to be adjusted accordingly.

12. Adding fluorescent beads of a different color without incubation causes these beads to concentrate in protein conjugated areas and facilitate the observation of pattern. Their use is optional. Alternatively the micropattern may be visualized using fluorescently labeled ECM protein.

13. The time for placing the stamped coverslip to the acrylamide solution is approximately 4 min after it is mixed with the first fluorescent beads. We have found that this overall incubation time works well for the current condition (*see* **Note 11**).

14. With this method, fluorescent beads with pre-incubation show a homogeneous distribution (Fig. 2a, b), while the beads without incubation reveal the micropattern (Fig. 2c, d).

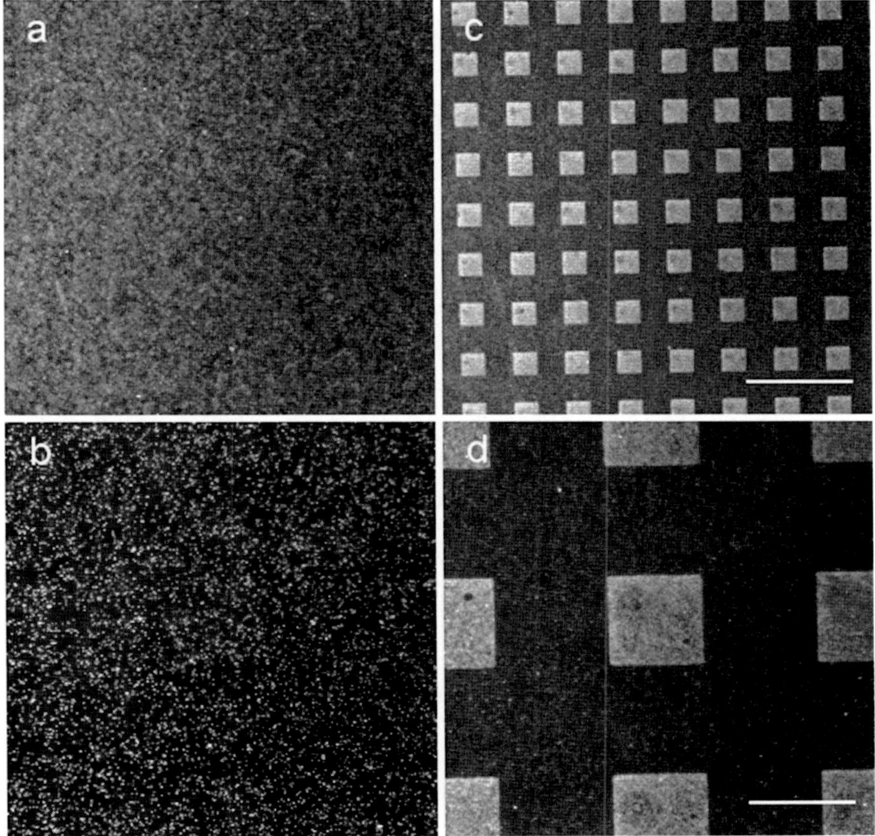

Fig. 2 Distribution of fluorescent beads embedded in the polyacrylamide hydrogel. Beads pre-incubated with protein show homogeneous distribution, suitable for traction force microscopy (**a, b**). In contrast, beads added immediately before gel polymerization concentrate in ECM-patterned areas, suitable for visualizing the pattern (**c, d**). Scale bar, 200 μm (**a, c**), 50 μm (**b, d**)

Acknowledgement

This work was supported by grant GM-32476 from the National Institutes of Health to Y.L.W.

References

1. Tambe DT, Corey-Hardin C, Angelini TE et al (2011) Collective cell guidance by cooperative intercellular forces. Nat Mater 10:469–475

2. Janmey PA, Miller RT (2011) Mechanisms of mechanical signaling in development and disease. J Cell Sci 124:9–18

3. Chen CS, Mrksich M, Huang S et al (1997) Geometric control of cell life and death. Science 276:1425–1428

4. Singhvi R, Kumar A, Lopez G et al (1994) Engineering cell shape and function. Science 264:696–698

5. Dike L, Chen C, Mrksich M et al (1999) Geometric control of switching between growth, apoptosis, and differentiation during angiogenesis using micropatterned substrates. In Vitro Cell Dev Biol Anim 35:441–448

6. McBeath R, Pirone DM, Nelson CM et al (2004) Cell shape, cytoskeletal tension, and rhoa regulate stem cell lineage commitment. Dev Cell 6:483–495

7. Kilian KA, Bugarija B, Lahn BT et al (2010) Geometric cues for directing the differentiation of mesenchymal stem cells. Proc Natl Acad Sci 107:4872–4877

8. Wang H-B, Dembo M, Wang Y-L (2000) Substrate flexibility regulates growth and apoptosis of normal but not transformed cells. Am J Physiol Cell Physiol 279:C1345–C1350

9. Engler AJ, Sen S, Sweeney HL et al (2006) Matrix elasticity directs stem cell lineage specification. Cell 126:677–689

10. Pelham RJ, Wang Y-l (1997) Cell locomotion and focal adhesions are regulated by substrate flexibility. Proc Natl Acad Sci 94:13661–13665

11. Rape A, Guo W-H, Wang Y-L (2011) Responses of cells to adhesion-mediated signals: a universal mechanism. In: Wagoner JA, Harley BAC (eds) Mechanobiology of cell-cell and cell-matrix interactions. Springer, USA, pp 1–10

12. Wang Y-L, Pelham RJ Jr (1998) Preparation of a flexible, porous polyacrylamide substrate for mechanical studies of cultured cells. In: Richard BV (ed) Methods in enzymology. Academic, New York, pp 489–496

13. Damljanovic V, Lagerholm BC, Jacobson K (2005) Bulk and micropatterned conjugation of extracellular matrix proteins to characterized polyacrylamide substrates for cell mechanotransduction assays. Biotechniques 39:847–851

14. Guo W-h, Wang Y-I (2007) Retrograde fluxes of focal adhesion proteins in response to cell migration and mechanical signals. Mol Biol Cell 18:4519–4527

15. Wang N, Ostuni E, Whitesides GM et al (2002) Micropatterning tractional forces in living cells. Cell Motil Cytoskeleton 52:97–106

16. Vignaud T, Galland R, Tseng Q et al (2012) Reprogramming cell shape with laser nanopatterning. J Cell Sci 125:2134–2140

17. Rape AD, Guo W-h, Wang Y-l (2011) The regulation of traction force in relation to cell shape and focal adhesions. Biomaterials 32:2043–2051

18. Fischer B, Thomas KB, Dorner F (1998) Collagen covalently immobilized onto plastic surfaces simplifies measurement of von Willebrand factor-collagen binding activity. Ann Hematol 76:159–166

19. Li F, Li B, Wang Q-M et al (2008) Cell shape regulates collagen type I expression in human tendon fibroblasts. Cell Motil Cytoskeleton 65:332–341

20. Reinhart-King CA, Dembo M, Hammer DA (2003) Endothelial cell traction forces on RGD-derivatized polyacrylamide substrata. Langmuir 19:1573–1579

21. Guo W-H, Wang Y-l (2011) Micropatterning cell-substrate adhesions using linear polyacrylamide as the blocking agent. Cold Spring Harb Protoc 2011(3):prot5582

Chapter 14

Measuring Cell–Cell Tugging Forces Using Bowtie-Patterned mPADs (Microarray Post Detectors)

Daniel M. Cohen, Mike T. Yang, and Christopher S. Chen

Abstract

Cells generate traction forces upon adhesion to the extracellular matrix as well as to neighboring cells. These forces are important for the growth and maintenance of adhesion structures such as focal adhesions and adherens junctions, and may play roles in tissue development. Here, we describe a method for measuring the tugging force transmitted across the cell–cell junction between two paired cells.

Key words Cell–cell tugging force, Cell–cell junctions, Micropatterning, mPADs, Traction force

1 Introduction

The ability of cell–matrix and cell–cell adhesions to support transmission of mechanical forces (either cell-generated or externally applied) has been demonstrated through a variety of methods including traction force microscopy [1–4], atomic or molecular force microscopy [5–7], optical or magnetic bead traps [8–11], and molecular tension reporters [12–15]. These methods have been especially useful in demonstrating that cells develop forces in response to newly formed adhesions, as well as in testing how adhesions respond/remodel upon loading with externally applied forces. However, measurements of mechanical forces that are transmitted across endogenous cell–cell junctions, in particular, have eluded these tools. Here, we present a general strategy for measuring cell–cell tugging forces between pairs of cells. This approach has been successfully applied to measuring how cell–cell tugging forces between endothelial cells respond to vasoactive compounds, as well as to Rho and Rac GTPase signaling agonists/antagonists [16].

Our method employs culturing cells on specialized substrates, mPADs, comprised of elastomeric posts whose deformations quantitatively report cellular traction forces [4, 16]. These mPAD arrays

Troy A. Baudino (ed.), *Cell-Cell Interactions: Methods and Protocols*, Methods in Molecular Biology, vol. 1066,
DOI 10.1007/978-1-62703-604-7_14, © Springer Science+Business Media New York 2013

are available as a resource to the broader research community upon request from the Chen lab. To increase both the yield of cell–cell contact and promote more uniform cell–cell adhesions, the mPADs are functionalized with bowtie-shaped islands of fibronectin [17]. In these patterns, cell–cell contact is geometrically biased towards the neck of the bowtie, whereas the limited adhesion area per bowtie selects against the formation of multicellular (>2 cell) clusters. In this configuration, the tugging force exerted by one cell is counterbalanced by the traction forces in the adjacent, contacting cell (net forces are in equilibrium, and therefore sum to zero). Thus the summation of traction forces under one cell yields a resultant force vector that is equal in magnitude and opposite in direction to the cell–cell tugging force exerted by its neighbor.

Using this method, we have previously demonstrated that endothelial cells generate cell–cell tugging forces on the order of 40 nN, with an apparent stress of approximately 1 nN/μm^2 at sites of cell–cell contact [16]. Moreover, this tugging force was required for the maintenance of the adherens junction, a finding that has now been corroborated by experiments correlating force, myosin, and cell–cell junction assembly in several cell types [18–22]. The mPAD system is amenable to studying cell–cell tugging forces in a variety of adherent cell types and can be easily adapted to either end-point analyses to measurement of dynamic force responses to extracellular stimuli [16, 23].

2 Materials

2.1 *Bowtie Stamping and mPAD Reagents*

1. Sylgard 184 polydimethylsiloxane (PDMS; Dow Corning, Morgan Hill, CA, USA).

2. Bowtie-patterned silicon wafers, as well as mPAD arrays can be generated by end-users with access to photolithography and microfabrication capabilities. Alternatively, mPADs are available upon request from the Chen Lab.

3. Prepare 5 mg/ml fibronectin (BD Biosciences, San Jose, CA, USA) by dissolving 5 mg of fibronectin in 1 ml of sterile water. Aliquot and freeze at –20 °C.

4. Prepare 2 % F127 Pluronics (Sigma-Aldrich, St. Louis, MO, USA) stock solution by dissolving 2 g of F127 Pluronics in 100 ml of distilled water in a bottle top 0.22 μm filter. Allow Pluronics to dissolve and filter by gravity flow overnight. Dilute 1:10 into phosphate buffered saline (PBS) for a 0.2 % working solution.

5. MatTek Petri dishes (35 mm, with 20 mm holes; MatTek Corporation, Ashland, MA, USA).

6. Prepare DiI stock solution by dissolving 25 mg of DiI (1,1′-dioleyl-3,3,3′,3′-tetramethylindocarbocyanine methane sulfonate) (Life Technologies, Grand Island, NY, USA) in 500 ml ethanol. Remove aggregates using a 0.22 μm filter. Store at 4 °C in a dark or foil-wrapped bottle to protect from light.

2.2 Immunostaining Reagents

1. Add 10 ml of 16 % paraformaldehyde to 26 ml of distilled water, and 4 ml of 10× PBS to make a 4 % solution. Add 100 ml of 10 % (v/v) Triton X-100 to 10 ml of 4 % paraformaldehyde to make permeablilization/fixation solution.

2. Dissolve 5 ml of goat serum (Life Technologies) in 45 ml of PBS to make 10 % (v/v) blocking buffer.

3. Anti-β-catenin (BD Biosciences) is diluted 1:100 in 10 % blocking buffer.

4. AlexaFluor 647 goat anti-mouse secondary antibody (Life Technologies) is provided as a 2 mg/ml solution. Use at 1:400 to 1:800 in 10 % blocking buffer.

5. Dissolve AlexaFluor 488 phalloidin (Life Technologies) in 1.5 ml of methanol. Dilute 1:100 in 10 % blocking buffer. Dissolve 10 mg of DAPI in 2 ml of water to yield 5 mg/ml stock solution. Use at 1:2,000 dilution in 10 % blocking buffer for immunostaining.

6. Fluoromount G (Electron Microscopy Sciences).

---

3   Methods

3.1 Generation of mPAD Substrates

For the purposes of simplicity, herein we provide a brief overview of the generation of silicon wafers that serve as templates for mPAD arrays. Design and fabrication of novel mPAD arrays requires users to have access to advanced microfabrication facilities and interested readers are referred elsewhere for detailed protocols [24–26].

Briefly, a photomask is generated containing the micropillar features in a regular square or hexagonal array. This photomask is used to pattern SU-8 photoresist that has been spin-coated on to a silicon wafer. Using a mask aligner and UV light source, the SU-8 becomes crosslinked by UV light in regions that are not protected by the photomask. Noncrosslinked SU-8 is removed, leaving behind micropillars of 7–12 μm height and a 3–6 μm diameter (specified by the features in the photomask). High-density mPAD arrays comprised of 2 μm posts, with 4 μm post-to-post spacing, have been generated using SU-8 photopatterning, but they require a more sophisticated process [24]. These silicon masters are used to cast negative PDMS molds (where the features are recessed wells). These negative PDMS molds are then used to cast

PDMS-based replicas of the micropillar substrates; herein referred to as mPAD substrates.

Silicon wafers for bowtie patterns use a similar SU-8 photopatterning approach, but do not require the generation of negative PDMS molds. Instead, the silicon wafers contain recessed wells in the shape of bowties, and PDMS stamps that contain the bowtie patterns are cast directly from the bowtie master (Subheading 3.2).

3.2 Functionalization of mPADs with Bowtie Micropatterns

1. Prepare PDMS stamps as follows. Mix 60 g PDMS with 2 g curing agent in a plastic cup. Stir vigorously with a pipette for 3–5 min.

2. Degas PDMS solution in a vacuum desiccator until air bubbles are completely removed (30–60 min).

3. Pour a 6–7 mm thickness layer of PDMS over a silicon wafer containing bowtie-shaped micropatterns in an aluminum dish (*see* **Note 1**). Make sure bubbles introduced while pouring the PDMS dissipate before proceeding to next step.

4. Bake stamps for 15 min at 110 °C to cure the PDMS. Cool on bench for approximately 5 min.

5. Gently peel PDMS stamps from silicon wafer and trim away excess PDMS. Cut stamps into squares sized-matched to the mPAD substrates (typically 21×21 mm) (*see* **Note 2**). Using a razor blade, notch the nonfeatured side of the stamp to indicate the bottom. Sterilize stamps by briefly (2–5 min) rinsing in 70 % ethanol.

6. Dilute fibronectin to 50 μg/ml in sterile water (*see* **Note 3**). In a cell culture hood, place the stamps in a deep petri dish. Add the fibronectin solution in a dropwise fashion to the top (featured side) of the PDMS stamp. Gently drag droplets using a micropipettor until drops merge into a continuous surface covering the entire stamp. 50–100 μl of solution is needed to cover each stamp. Coat stamps for 60 min at room temperature.

7. Flood stamps with sterile water. Pick up stamps with tweezers and rinse twice more in a bath of sterile water. Dry stamps completely with compressed nitrogen (air gun). Place stamps face up in a fresh petri dish.

8. Render mPAD substrates hydrophilic by UV-ozone treatment in a UVO cleaner (model no. 342, Jelight, Irvine, CA, USA) for 7 min.

9. Transfer mPADs from the UVO cleaner into a tissue culture hood. Position bowtie stamps onto mPADs using tweezers. Align stamp with one edge of the mPAD and allow stamp to slowly fall onto the substrate. Conformal contact can be verified by visual inspection of the substrates. From the perspective of an acute angle, a diffraction pattern should appear at the stamp:mPAD interface. If an incomplete pattern is present,

apply gentle pressure on the stamp using tweezers until conformal contact is achieved. Allow 15 s for protein transfer.

10. Submerge mPADs in 1.4 ml of 100 % ethanol (mPADs should either be attached to a Mattek dish for live cell imaging or placed in a 35 mm petri dish). Displace stamp from mPADs with a gently knocking motion (do not peel the stamp off as this can collapse the mPADs). Set stamps aside (*see* **Note 4**).

11. Add 0.6 ml of water to the mPAD dish and swirl to mix. Remove this 70 % ethanol solution by aspiration and rinse 3× with sterile milliQ water. Do not allow the substrates to dewet (keep samples flat, not tilted, during aspiration steps).

12. Label the PDMS with DiI. Dilute a 50 µg/ml DiI stock solution to 5 µg/ml working concentration using sterile milliQ water (2 ml per substrate). Add to mPADs and incubate for 60 min. Cover sample with aluminum foil to protect from light in all subsequent steps.

13. Aspirate DiI solution and rinse 3× with milliQ water. Add 0.2 % Pluronics F-127 to passivate the nonfunctionalized regions of PDMS. Allow 30–60 min for adsorption of Pluronics.

14. Rinse 3× with milliQ water. mPADs can now be used for cell seeding or stored in PBS at 4 °C for up to 1 week until cells are ready.

3.3 Seeding mPADs with Cells

1. Aspirate PBS or water from mPADs and replace with 2 ml of cell culture medium.

2. Trypsinize and resuspend cells to 20,000–100,000 cells/ml. Add 0.5 ml of cell suspension to the mPADs (to achieve a seeding density of 1,000–5,000 cells/cm^2). Be sure to disrupt or remove any clumps of cells to avoid occupancy of bowties by more than two cells.

3. Allow cells to adhere to mPAD substrates for 10 min in a tissue culture incubator (*see* **Note 5**).

4. Rinse away excess nonadherent cells. If mPADs are attached to a Mattek dish, aspirate and replenish medium 3× or until no floating cells are observed. Alternatively, mPADs on glass coverslips can be successively transferred to new 35 mm dishes with fresh culture medium to remove nonadherent cells.

5. Return mPADs to tissue culture incubator and allow the cells to spread overnight.

6. If desired, change medium per experimental needs. Serum starvation (0.1–0.5 % serum) is recommended to observe changes in traction and tugging forces in response to the addition of soluble agonists. Serum starvation periods of 18–24 h are usually adequate to reduce cellular contractility to basal levels.

7. Upon stimulation, changes in cell traction and tugging forces occur within 2–30 min. Treat cells with the manipulation or stimulus of interest (or a vehicle control), rinse briefly with PBS, and fix at the desired end-point.

8. Fix/permeabilize cells with 2 ml of room temperature 4 % paraformaldehyde containing 0.1 % Triton X-100 for 3 min, followed by an additional fixation in 4 % paraformaldehyde for 15 min (*see* **Note 6**). If multiple conditions are being tested in parallel, be sure to stagger the fixation times. Force measurements are quite sensitive to variations in rinsing and fixation times. Alternatively, time lapse imaging of mPAD displacements can be performed during this time (*see* **Note 7**).

3.4 Immunostaining

1. Aspirate paraformaldehyde, and rinse for 5 min with PBS. Prepare samples for immunofluorescence by blocking with 10 % goat serum (in PBS) or an equivalent blocking agent.

2. Immunostain for cell–cell junctions using an appropriate antibody. We recommend mouse anti-β-catenin at a 1:100 dilution in blocking buffer (10 % goat serum). For antibody incubations, invert mPADs onto a 200 μl droplet of diluted primary antibody on a piece of parafilm. Be sure to avoid contacting mPADs directly with tweezers (handle edge of coverslip only) to avoid collapse of the posts. Incubate mPADs with primary antibody for 1 h at room temperature.

3. Place mPAD substrates in a 35 mm dish and rinse 4× for 5 min with PBS.

4. Immunostain mPADs with secondary antibody for 1 h at room temperature as in **step 2**. Cover samples to protect from light.

5. Place mPAD substrates in a 35 mm dish and rinse 4× for 5 min with PBS.

6. Optional: Stain cells with fluorescent phalloidin to visualize the actin cytoskeleton and determine cell boundaries during mPAD quantification. Counterstain nuclei with DAPI. Stain for 20 min in a 200 μl droplet as in **step 4** (if desired, this step can be combined with **step 4**, as extending the incubation to >20 min is not problematic). Cover samples to protect from light.

7. Place mPAD substrates in a 35 mm dish and rinse 4× for 5 min with PBS.

8. Mount mPADs on slide using Fluoromount G. Use a small drop to adhere mPAD face-up to glass slide. Cover top of mPADs with a few drops of Fluoromount G and seal with a fresh 22 mm glass coverslip. Allow Fluoromount to cure for 4 h. Fluorescent microscopy imaging should be done with an inverted microscope.

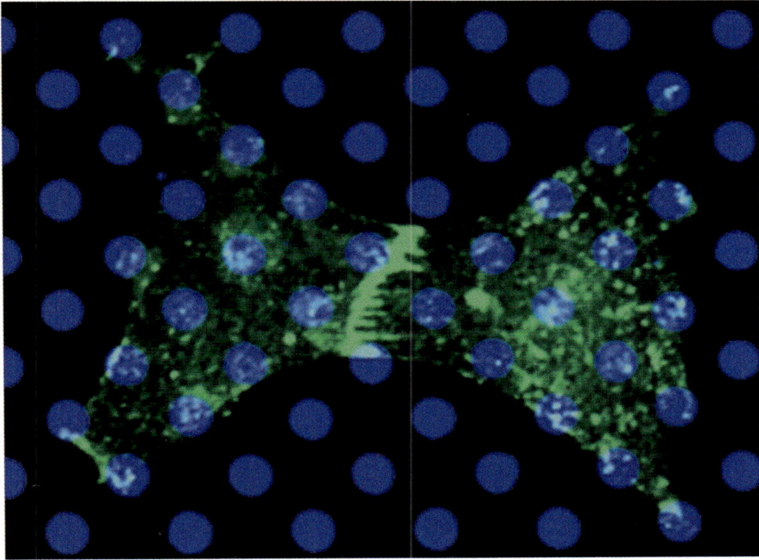

Fig. 1 Cell–cell junctional staining of bowtie-patterned endothelial cells on mPADs. DiI-stained microposts are shown in *blue* and β-catenin immunostaining shown in *green*. Note cell–cell junction in center bowtie. Diffuse cytoplasmic β-catenin in cell body is removed by thresholding in subsequent image processing steps

3.5 Image Acquisition and Quantification of Cell–Cell Junctions

1. Using an inverted microscope, collect multi-channel images to define the position of the cell nucleus (DAPI stain), the cell boundaries (phalloidin, green channel), and the cell–cell junctions (anti-β-catenin, far red/infrared channel). Oil immersion objectives are recommended with a magnification of at 40–63×. Microscopy can be done either in epifluorescence or confocal modes (*see* **Notes 8** and **9**). An example of typical staining pattern of β-catenin overlaid on the micropost channel is provided in Fig. 1.

2. Acquire images of microposts tips using the DiI signal (red channel); henceforth the "top image". Make sure that the image includes two peripheral rows of posts surrounding the bowtie region (*see* **Note 10**). It is best to collect at least 20–30 sets of images of bowties in **steps 1** and **2**.

3. Quantify cell–cell junction size and brightness using appropriate image analysis software (e.g. Matlab, MetaMorph). For epifluorescence images, perform a flatfield correction to correct for inhomogeneities in illumination, followed by pixels thresholding to select the brightest 25 % of pixels in the β-catenin channel. The resultant image is then binarized and all pixels are summed to generate an area measurement for cell–cell junctions. In addition, mean pixel intensity can be determined by averaging the intensity values of the β-catenin pixels in the thresholded image.

A MATLAB-based program for traction force measurements is available upon request from the Chen Lab. Here we simply describe the general approach and the extra steps required for determination of cell tugging force.

1. Although the mPADs conform to a regular array (either square or hexagonally packed), this array may be rotated arbitrarily in any given image due to variability in how the sample was mounted relative to the camera's field of view on the microscope. Images should be rotated to match a single Cartesian coordinate system, such that the rows of microposts align with the horizontal X-axis and vertical Y-axis.

2. Generate an ideal grid to predict the expected location for all posts using linear interpolation. In each row of microposts, "guide posts" are selected from the left and right side of the bowtie. Using these guide posts, an equation is determined for the best-fit line through the center of the posts. Since the spacing between posts is fixed, the number of posts, and position of these posts between the guide posts can be determined by the post-to-post spacing constant (*see* **Note 11**).

3. Determine which posts are in contact with a cell. Overlay the phalloidin and/or β-catenin images on the mPAD (DiI) image. If desired, a registration step can be done to correct for translational shifts between the images acquired in different fluorescence channels. All posts positioned under the cell body or associated with cellular protrusions such as filopodia are considered to be in contact with a cell.

4. Find the centroid of the posts and determine deflections. Post positions are determined by thresholding the DiI fluorescence image. An appropriate threshold value will yield pixel segments that approximate the cross-sectional area of the posts. At this point, the centroid can be determined for segmented group of pixels (i.e. each post). Subtract the experimentally determined centroid from the undeflected centroid location as determined by the ideal grid (Subheading 3.5, **step 2**) to calculate displacements per post. An illustration of the traction force vectors mapped on to the mPAD micrographs is provided in Fig. 2a.

5. Vectorial displacements can be converted to traction force vectors using the spring constant for the posts (*see* **Note 12**). From the traction force vectors, the magnitude and direction of forces are known. To calculate tugging force, all of the traction force vectors under a given cell are summed (Fig. 2b). In the case of an isolated cell, the X and Y components of force should sum to zero (equilibrium state). However, in the bowtie, the sum of the traction forces under each cell should be summed with the tugging force to achieve the equilibrium

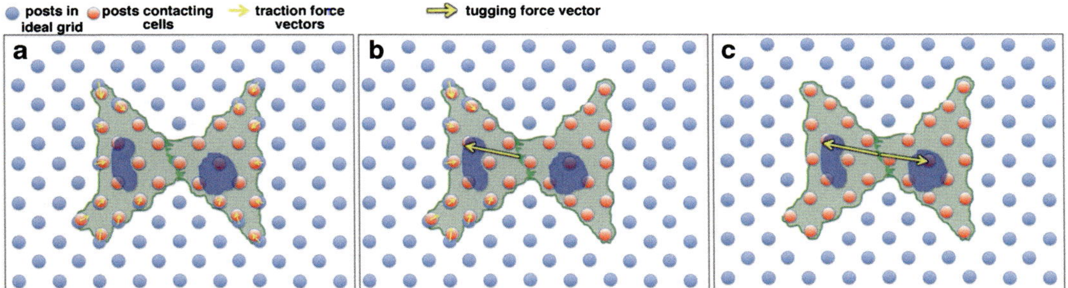

Fig. 2 Determination of cell–cell tugging force in bowtie-patterned cells on mPADs. (**a**) Top-down view of cells on mPADs. Each micropost is represented by a *circle*, showing either its position in the ideal grid (*blue*) versus its experimentally measured position (*red*). For simplicity, experimentally measured positions are limited here to those posts that are in contact with bowtie-patterned cells. The distance between centroids of the *blue* versus *red* posts is used to calculate traction force vectors (*yellow arrows*) using the spring constant of the posts (*see* **Note 12**). (**b**) Determination of tugging force vector for the cell located in the *left half* of the bowtie. Considering *only* the traction forces in the *left cell*, the *x* and components of the vectors are summed to yield a resultant traction force vector. The tugging force vector is then plotted as the *equal*, and *opposite*, force vector such that the net forces sum to zero. (**c**) The process in panel **b** is repeated for the cell located in the *right half* of the bowtie, yielding an independent tugging force vector for each cell. Tugging force per bowtie is reported as the average magnitude of the two tugging force vectors

state. Therefore the tugging force is calculated as an equal and opposite force to the resultant vector for all traction forces in the neighboring cell (*see* **Note 13**). This process is repeated for each cell in the bowtie pair to generate two independent tugging force vectors (Fig. 2c). Average tugging force per bowtie is reported as the average magnitude of the two tugging force vectors.

4   Notes

1. We strongly suggest the use of bowtie patterns to enhance the yield of paired cells with well-defined cell–cell contacts (other geometric patterns, or featureless, flat stamps can be used in principle). Some optimization may be needed to select an appropriate bowtie pattern for a given cell type; ideally, cell spreading in each half of the bowtie should approximate the average area occupied by a cell when grown to confluency (typically 40–50 % of the area of maximal cell spreading observed in subconfluent conditions). Bowtie geometries are comprised of two isosceles triangles (2:2:3 dimensional ratio) that join at their vertices. Upon transfer of the bowtie patterns to the mPAD surfaces, the gap of nonadhesive surface that cells need to span in order to form cell–cell contacts will depending on micropost spacing (typically on the order of 4–9 μm).

2. For a detailed demonstration of generating and using PDMS stamps, readers are referred to ref. 27.

3. Fibronectin solutions need not be made freshly every time; however, dilute fibronectin solutions denature over time and should not be kept more than 2–3 weeks at 4 °C. Other matrix proteins such as collagen are compatible with microcontact printing, and can be used in lieu of fibronectin.

4. PDMS stamps can be reused in future experiments. After each use, sonicate for 5 min in 95 % ethanol to remove any residual fibronectin from the stamps.

5. Timing for cell adhesion to mPAD substrates can be dependent on several factors (e.g. cell type, medium composition, type of extracellular matrix in bowties, quality of stamping). If cells are not attached by 10 min, monitor the attachment in 5 min intervals until at least half the patterns show cell attachment. If excessive pooling of cells occurs, gently rock samples to disperse cells evenly.

6. Simultaneous permeabilization and fixation is recommended to remove the diffuse cytoplasmic pool of β-catenin that can otherwise obscure imaging of cell–cell junctions. This permeabilization step does not appear to interfere with the contractility state of the cell.

7. Live cell microscopy, when available, offers several distinct advantages to end-point based analyses for measurement of tugging forces; it can avoid artifacts associated with changes in contractility during the fixation process and enable insight into dynamics of tugging forces. However, tugging force analysis on live cells requires a vital label for cell–cell junctions. For example, in endothelial cells, adenoviral delivery of GFP-VE-cadherin was used to visualize the junctions [28]. Choice of label and mode of delivery need to be optimized for the cell type of interest.

8. If available, confocal microscopy is especially useful for imaging the cell–cell junctions by eliminating out-of-plane fluorescence. However, we find that it is possible to compensate for shortcomings of epifluorescence using image processing described here. Access to confocal imaging obviates the need for these processing steps.

9. When collecting images, adjust the z-position to select the appropriate focus position for each fluorescent marker (be sure to capture a phalloidin image near the base of the cell to determine which microposts are in contact with a cell). Be sure to select bowties that have exactly two cells and show a distinct

cell–cell junction at or near the neck of the bowtie. It is critical to use a constant exposure time for the β-catenin and DiI channels or subsequent quantitative analyses will not be possible.

10. The force detection algorithm requires information about nondeflected posts. Therefore, it is critical to select a magnification in which this top image contains at least two rows of nondeflected posts surrounding the bowtie-patterned cells. If necessary, switch to a lower magnification such that this requirement is met. The optimal focal plane for top images can be found by focusing down from the plane of the cell until the peripheral microposts (non-cell contacting) come sharply into view (post-tips should be perfectly circular, with the notable exception of severely deflected posts under a cell).

11. This calculation requires prior knowledge of the scale size for each image (i.e. how many microns per pixel). In principle, linear interpolation can be applying to incorporate positional information on both the horizontal and vertical axes; however, this approach does not significantly increase accuracy of the ideal grid and may even be less robust in cases where uneven illumination or other image distortions introduce a systematic error in micropillar position across the field of view.

12. Forces can be predicted by Hooke's law; $F = k \times X$, where k represents the spring constant and X is the measured displacement. The spring constant of the post is determined by the equation, $k = 3 \times E \times I / L^3$, where E is the elastic modulus, I is the area moment of inertia, and L is the length of the post. Spring constants of mPADs have also been empirically determined using micromanipulators to apply known forces to the posts.

13. The inference that the tugging force acts as a balancing force for the summed traction forces is only valid for pairs of cells. If more than two cells are in contact, it is impossible to know how the balancing forces imparted by the cell–cell tugging forces are distributed between them (there is no unique solution to the force balance equation). Therefore, this method cannot be generalized to multicellular clusters.

Acknowledgements

This work was supported by NIH grants HL73305 and through the RESBIO program (Integrated Technology Resource for Polymeric Biomaterials, NIH grant EB001046).

References

1. Lee J, Leonard M, Oliver T (1994) Traction forces generated by locomoting keratocytes. J Cell Biol 127:1957–1964

2. Pelham RJ, Wang Y (1999) High resolution detection of mechanical forces exerted by locomoting fibroblasts on the substrate. Mol Biol Cell 10:935–945

3. Balaban NQ, Schwarz US, Riveline D et al (2001) Force and focal adhesion assembly: a close relationship studied using elastic micropatterned substrates. Nat Cell Biol 3:466–472

4. Tan JL, Tien J, Pirone DM et al (2003) Cells lying on a bed of microneedles: an approach to isolate mechanical force. Proc Natl Acad Sci USA 100:1484–1489

5. Perret E, Leung A, Feracci H et al (2004) Trans-bonded pairs of E-cadherin exhibit a remarkable hierarchy of mechanical strengths. Proc Natl Acad Sci USA 101:16472–16477

6. Baumgartner W, Hinterdorfer P, Ness W et al (2009) Cadherin interaction probed by atomic force microscopy. Proc Natl Acad Sci USA 97:4005–4010

7. Panorchan P, Thompson MS, Davis KJ et al (2006) Single-molecule analysis of cadherin-mediated cell-cell adhesion. J Cell Sci 119:66–74

8. Potard US, Butler JP, Wang N (1997) Cytoskeletal mechanics in confluent epithelial cells probed through integrins and E-cadherins. Am J Physiol 272:C1654–1663

9. Ko KS, Arora PD, McCulloch CA (2001) Cadherins mediate intercellular mechanical signaling in fibroblasts by activation of stretch-sensitive calcium-permeable channels. J Biol Chem 276:35967–35977

10. Felsenfeld DP, Choquet D, Sheetz MP (1996) Ligand binding regulates the directed movement of beta1 integrins on fibroblasts. Nature 383:438–440

11. Galbraith CG, Yamada KM, Sheetz MP (2002) The relationship between force and focal complex development. J Cell Biol 159:695–705

12. Ohashi T, Kiehart DP, Erickson HP (2002) Dual labeling of the fibronectin matrix and actin cytoskeleton with green fluorescent protein variants. J Cell Sci 115:1221–1229

13. Smith ML, Gourdon D, Little WC et al (2007) Force-induced unfolding of fibronectin in the extracellular matrix of living cells. PLoS Biol 5:e268

14. Kong HJ, Polte TR, Alsberg E et al (2005) FRET measurements of cell-traction forces and nano-scale clustering of adhesion ligands varied by substrate stiffness. Proc Natl Acad Sci USA 102:4300–4305

15. Grashoff C, Hoffman BD, Brenner MD et al (2010) Measuring mechanical tension across vinculin reveals regulation of focal adhesion dynamics. Nature 466:263–266

16. Liu Z, Tan JL, Cohen DM et al (2010) Mechanical tugging force regulates the size of cell-cell junctions. Proc Natl Acad Sci USA 107:9944–9949

17. Nelson CM, Liu WF, Chen CS (2007) Manipulation of cell-cell adhesion using bowtie-shaped microwells. Methods Mol Biol 370:1–10

18. Maruthamuthu V, Sabass B, Schwarz US et al (2011) Cell-ECM traction force modulates endogenous tension at cell-cell contacts. Proc Natl Acad Sci USA 108:4708–4713

19. Miyake Y, Inoue N, Nishimura K et al (2006) Actomyosin tension is required for correct recruitment of adherens junction components and zonula occludens formation. Exp Cell Res 312:1637–1650

20. Shewan AM, Maddugoda M, Kraemer A et al (2005) Myosin 2 is a key Rho kinase target necessary for the local concentration of E-cadherin at cell-cell contacts. Mol Biol Cell 16:4531–4542

21. Huveneers S, Oldenburg J, Spanjaard E et al (2012) Vinculin associates with endothelial VE-cadherin junctions to control force-dependent remodeling. J Cell Biol 196:641–652

22. Yonemura S, Wada Y, Watanabe T et al (2010) alpha-Catenin as a tension transducer that induces adherens junction development. Nat Cell Biol 12:533–542

23. Yang MT, Reich DH, Chen CS (2011) Measurement and analysis of traction force dynamics in response to vasoactive agonists. Integr Biol (Camb) 3:663–674

24. Yang MT, Fu J, Wang YK et al (2011) Assaying stem cell mechanobiology on microfabricated elastomeric substrates with geometrically modulated rigidity. Nat Protoc 6:187–213

25. Sniadecki NJ, Chen CS (2007) Microfabricated silicone elastomeric post arrays for measuring traction forces of adherent cells. Methods Cell Biol 83:313–328

26. Desai RA, Yang MT, Sniadecki NJ et al (2007) Microfabricated post-array-detectors (mPADs): an approach to isolate mechanical forces. J Vis Exp 8:e311

27. Shen K, Qi J, Kam LC (2008) Microcontact printing of proteins for cell biology. J Vis Exp 22:e1065

28. Shaw SK, Bamba PS, Perkins BN et al (2001) Real-time imaging of vascular endothelial-cadherin during leukocyte transmigration across endothelium. J Immunol 167:2323–2330

Chapter 15

Generation and Analysis of Biosensors to Measure Mechanical Forces Within Cells

Katharina Austen, Carleen Kluger, Andrea Freikamp, Anna Chrostek-Grashoff, and Carsten Grashoff

Abstract

The inability to measure mechanical forces within cells has been limiting our understanding of how mechanical information is processed on the molecular level. In this chapter, we describe a method that allows the analysis of force propagation across distinct proteins within living cells using Förster resonance energy transfer (FRET)-based biosensors.

Key words Mechanobiology, Mechanotransduction, Force measurement, Biosensor, Tension sensor, Förster resonance energy transfer (FRET), Fluorescence-lifetime imaging microscopy (FLIM)

---

1   Introduction

The ability of cells to sense and respond to mechanical forces is central to many developmental and physiological processes but is also involved in numerous diseases such as atherosclerosis, cardiomyopathies, muscular dystrophies, skin disorders, and cancer [1]. The molecular mechanisms underlying cellular mechanosensitivity, however, are largely unclear because often we do not know how mechanical forces, typically in the range of nanonewton (nN) to piconewton (pN), are processed in living cells. Thus, a number of techniques have been developed to quantify mechanical parameters in cell culture [2]: cell traction force microscopy (CTFM) on polyacrylamide gels or micro-patterned surfaces to measure intrinsically generated forces in the nN range [3], micropipette aspiration techniques [4] or optical trapping to investigate cells' mechanical properties, and atomic force microscopy methods (AFM) [5] as well as magnetic twisting cytometry (MTC) [6] to probe subcellular

---

Katharina Austen, Carleen Kluger, and Andrea Freikamp have contributed equally to this chapter.

Troy A. Baudino (ed.), *Cell-Cell Interactions: Methods and Protocols*, Methods in Molecular Biology, vol. 1066, DOI 10.1007/978-1-62703-604-7_15, © Springer Science+Business Media New York 2013

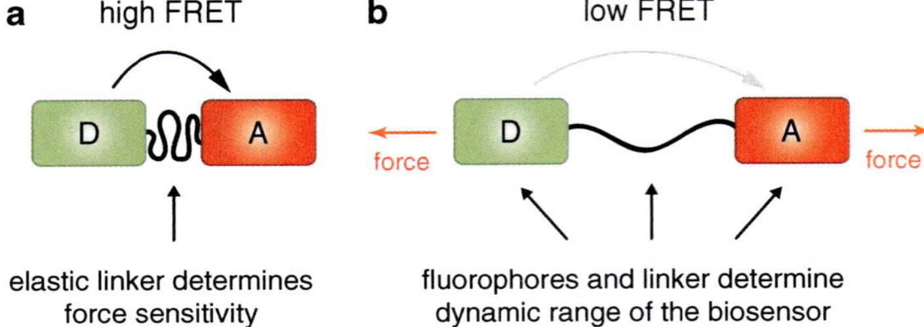

Fig. 1 Principle of a FRET-based tension sensor. (**a**) The tension sensor consists of a donor (D) and an acceptor (A) fluorophore separated by an elastic linker. (**b**) Mechanical force across the sensor stretches the elastic linker thereby increasing the fluorophore separation distance and reducing FRET. The mechanical properties of the linker determine the force sensitivity of the sensor, whereas the dynamic range is determined by the photophysical properties of the fluorophores and the linker

structures with force sensitivities of 10^{-11} to 10^{-12} N. Here, we describe a technique that allows for the measurement of mechanical forces across specific proteins with pN-sensitivity in living cells using a genetically encoded biosensor [7, 8].

The technique is based on an effect called Förster resonance energy transfer (FRET), a nonradiative energy transfer, which can occur between an excited donor fluorophore and an adjacent acceptor molecule [9]. The rate of energy transfer, the FRET efficiency (E), depends on the spectral properties of donor and acceptor molecule as well as their separation distance (r) and can be described by $E = R_0^6/(R_0^6 + r^6)$, where R_0 is the fluorophore separation distance at which energy transfer is 50 %.

A force-sensitive biosensor can be generated by linking two suitable fluorophores (FRET pair) with an elastic peptide that will be stretched under tension leading to a decrease in FRET [7]. The mechanical properties of the linker will determine the force sensitivity of the biosensor, and its dynamic range and sensitivity are governed by both linker and the FRET pair (Fig. 1). A previously used biosensor is based on an mTFP1–VenusA206K FRET pair and employs an elastic linker derived from the spider silk protein flagelliform, which behaves as an elastomeric spring and can be reversibly stretched by forces in the range of 1–6 pN [7]. Biosensors probing other force ranges could be generated by using elastic linkers with distinct mechanical properties.

To generate a biosensor that reports forces across a protein of interest, the tension sensor module has to be inserted into the molecule (Fig. 2a, b); the insertion, however, may affect protein function. Therefore, biochemical and structural information about the target protein should be critically reviewed to identify feasible insertion sites that are not linked to a relevant function of the molecule. Furthermore, the following control constructs should be generated (*see* Subheading 3.1) to test the functionality of the

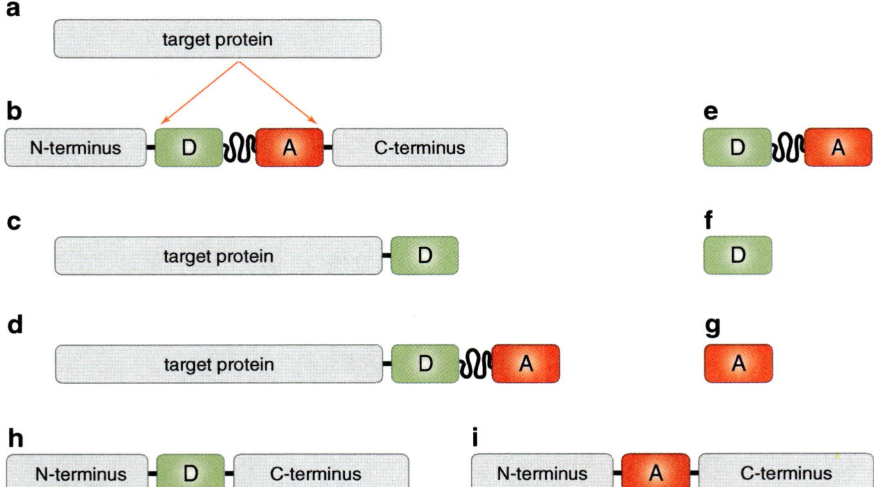

Fig. 2 Control and tension sensor constructs. (**a**) Schematic view of the protein of interest. (**b**) To generate the tension sensor construct the tension sensor module has to be inserted into the protein of interest. (**c**) The wt-control should behave as the endogenous protein and can be used to determine τ_D. (**d**) The zero-force FRET control can be used to determine τ_{DA} in the absence of tension as no physiologically relevant forces can be applied across the N or C terminal FRET module. (**e**) The tension sensor module, (**f**) the donor fluorophore, and (**g**) the acceptor fluorophore should be individually expressed in cells to analyze and evaluate FRET data. (**h, i**) Intermolecular FRET controls are generated by inserting the individual fluorophores into the protein of interest. Expression of both constructs within the same cell will allow an estimation of intermolecular FRET

modified protein and to control the FRET experiment: first, a wild-type (wt) form of the target protein tagged (N- or C-terminally) with the donor fluorophore (Fig. 2c, wt-control); second, a wt molecule tagged (N- or C-terminally) with the tension sensor module (Fig. 2d, zero-force FRET control); third, expression constructs of the tension sensor module, the donor and the acceptor fluorophore (Fig. 2e–g, FRET controls); fourth, two constructs, in which the donor or the acceptor fluorophore are inserted into the target protein at the position of tension sensor insertion (Fig. 2h, i, intermolecular FRET controls). All proteins should be transiently or stably expressed (*see* Subheading 3.2) in the cell line of choice and analyzed under comparable conditions.

Before FRET analysis the biosensor has to be tested for functionality. This evaluation can take many forms, depending on the protein of interest, but the following experiments are highly recommended. First, stable protein expression, at the expected molecular size and at appropriate expression levels, should be confirmed (*see* Subheading 3.3.1). Second, the subcellular localization of the biosensor as compared to the wt-control (Fig. 2c) should be tested (*see* Subheading 3.3.2). Finally, rescue experiments in cells, which have been depleted of the endogenous protein by RNA interference or gene targeting, should be performed to ensure that the biosensor can functionally replace the endogenous protein.

For some proteins this may require the generation of stable cell lines as described below (*see* Subheading 3.2.2).

Different ways to analyze FRET have been described [10]. Most laboratories rely on intensity-based methods, which can use standard wide-field or confocal microscopes. Intensity-based analyses, however, require rather complex post-acquisition processing methods, are prone to artifacts, and do not readily yield a quantitative measure of FRET efficiency. Even though technically more challenging, time-correlated single photon counting fluorescence-lifetime imaging microscopy (TCSPC-FLIM) [11, 12] avoids some of these difficulties and transfer rates can be directly obtained (*see* Subheadings 3.4 and 3.5).

TCSPC-FLIM requires a high-frequency pulsed laser and a detector able to record the arrival times of single photons with respect to the laser pulse. The resulting photon distribution can be used to calculate the fluorescence lifetime τ of a donor molecule in the presence (τ_{DA}) or absence (τ_{D}) of an acceptor. The FRET efficiency (E) is related to these lifetimes by: $E = 1 - (\tau_{DA}/\tau_{D})$.

Altogether, a precise design of the construct, a careful evaluation of the expressed biosensor and a quantitative data analysis will allow the correlation of FRET efficiencies to mechanical forces acting upon the protein of interest.

2 Materials

2.1 Generation of the Tension Sensor Expression Constructs

DNA oligonucleotides, plasmids, enzymes, and buffers should be stored at –20 °C.

1. TE buffer, pH 8.0: 10 mM Tris–HCl, 1 mM EDTA; diluted in ultrapure water.

2. DNA oligonucleotides, high-purity salt-free (HPSF): 100 μM stock solution in TE buffer, 10 μM working solution in ultrapure water.

3. Purified cDNA of the target protein.

4. PfuUltra II Fusion HS DNA Polymerase, 10× PfuUltra II Rxn buffer (Agilent Technologies, Santa Clara, CA, USA).

5. Deoxyribonucleotide triphosphate mix (dNTPs): 10 mM stock of each dNTP in ultrapure water.

6. Restriction enzymes, T4 DNA Ligase, and respective buffers.

7. pBluescript II Phagemid Vector (Agilent Technologies), pLPCX Vector (Clontech, Mountain View, CA, USA), pcDNA 3.1 (Life Technologies, Grand Island, NY, USA).

8. 1 kb DNA Ladder.

9. Software for sequence editing and virtual cloning such as Lasergene Core Suite (DNASTAR, Madison, WI, USA).

2.2 Expression of the Biosensor Constructs in Living Cells

2.2.1 Transient Expression

1. Phosphate-buffered saline (PBS), pH 7.4: 137 mM NaCl, 2.7 mM KCl, 10.1 mM Na_2HPO, 1.75 mM KH_2PO_4.

2. Transfection Reagent Lipofectamine 2000 (Life Technologies).

3. Opti-MEM Reduced Serum Medium with GlutaMAX (Life Technologies).

4. Growth medium: Dulbecco's Modified Eagle's Medium (DMEM) with high glucose, GlutaMAX, and pyruvate (Life Technologies) supplemented with 10 % Fetal Calf Serum (FCS) and 1× Penicillin/Streptomycin (P/S).

5. Transfection medium: growth medium without antibiotics.

6. 4 μg of purified expression plasmid containing the biosensor or the control cDNA expression construct.

2.2.2 Stable Expression

1. Gryphon packaging cell line (Allele Biotechnology, San Diego, CA, USA).

2. Growth medium: DMEM with high glucose, GlutaMAX, and pyruvate (Life Technologies) supplemented with 10 % FCS and 1× P/S.

3. 2× HBS buffer, pH 7.0: 280 mM NaCl, 50 mM HEPES, 1.5 mM Na_2HPO_4; sterile filtered.

4. 2 M $CaCl_2$ in ultrapure water; sterile filtered.

5. 20 μg purified expression plasmid containing the biosensor or the control cDNA expression construct.

6. Chloroquine diphosphate salt: 50 mM in PBS; sterile filtered, aliquots stored at −20 °C.

7. Polybrene (EMD Millipore, Billerica, MA, USA); aliquots stored at −20 °C.

8. Syringes (10 ml) and sterile filters (0.22 μm pore size; PES membrane).

2.3 Biosensor Evaluation

2.3.1 Biochemical Evaluation

1. Lysis buffer, pH 7.4: 50 mM Tris–HCl, 150 mM NaCl, 1 % Triton X-100 supplemented with complete ULTRA, Mini, EDTA-free protease, and PhosSTOP phosphatase inhibitor cocktail tablets (Roche Diagnostics, Indianapolis, IN, USA).

2. Loading buffer (4×), pH 6.8: 200 mM Tris–HCl, 4 mM EDTA, 84.5 % glycerol, 8 % sodium dodecyl sulfate (SDS), 4 % ®-mercaptoethanol, 0.05 % bromphenol blue.

3. BCA protein assay kit (Novagen, EMD Millipore, Billerica, MA, USA).

4. Immobilon-P transfer PVDF membrane (EMD Millipore).

5. Tris-buffered saline (TBS), pH 7.5: 150 mM NaCl, 50 mM Tris–HCl.

6. TBS-T: TBS with 0.1 % Tween 20.

7. Blocking solution: TBS-T with 5 % skim milk and 0.02 % NaN_3.

8. Primary antibody: polyclonal rabbit anti-GFP antibody (*see* **Note 1**).

9. Secondary antibody: polyclonal goat anti-rabbit IgG antibody conjugated to horseradish peroxidase (HRP).

10. Chemiluminescence detection kit: Immobilon Western (EMD Millipore).

11. Fujifilm LAS-4000 Luminescent image analyzer.

2.3.2 Immuno-histochemical Evaluation

1. Fibronectin (FN) from bovine plasma (Calbiochem, EMD Millipore) diluted to 5 µg/ml in PBS.

2. Glass coverslips No. 1.5, 12 mm.

3. 24-well plate.

4. 4 % paraformaldehyde (PFA) in PBS (*see* **Note 2**).

5. Blocking solution: PBS containing 2 % bovine serum albumin (BSA) and 0.1 % Triton X-100.

6. Microscope slides.

7. Mounting medium: ProLong Gold Antifade Reagent (Molecular Probes, Eugene, OR, USA).

8. Leica TCS SP5 X confocal laser scanning microscope (Leica Microsystems, Buffalo Grove, IL, USA).

9. HCX PL Apo 63× 1.4 NA, oil immersion objective (Leica Microsystems).

2.4 Live Cell Microscopy and Image Acquisition

1. PBS.

2. FN from bovine plasma (Calbiochem, EMD Millipore) diluted to 5 µg/ml in PBS.

3. UV light source for sterilization.

4. Live cell imaging medium: DMEM (without Phenol Red) supplemented with 4.5 g/l glucose, 25 mM HEPES, 2 mM glutamine, 10 % FCS and 1× P/S.

5. Leica TCS SP5 X confocal laser scanning microscope (Leica Microsystems) equipped with a pulsed white light laser (NKT Photonics, Morganville, NJ, USA).

6. FLIM X16 78 MHz TCSPC Detector (LaVision BioTec, Bielefeld, Germany).

7. X1 Port (Leica Microsystems).

8. HCX PL APO CS, 63× water objective (NA = 1.2) (Leica Microsystems).

9. Appropriate band pass filter with light transmission properties that depend on donor and acceptor fluorophores (*see* **Note 3**).

10. Software to operate scanning microscope: Leica LAS AF Software (Leica Microsystems).

11. Software to operate FLIM Detector: Imspector Software (LaVision BioTec).

2.5 Data Analysis and Evaluation

1. Software for post-acquisition data analysis: MATLAB (MathWorks, Natick, MA, USA).

2. LOCI Toolbox for MATLAB.

---

3 Methods

3.1 Generation of the Tension Sensor Expression Constructs

To generate the majority of control expression constructs (Fig. 2c–g), standard cloning techniques can be used that may vary between labs and will not be described here. However, we highly recommend the use of virtual cloning software such as Lasergene (DNASTAR) to avoid mistakes in the cloning strategy of the constructs and the design of oligonucleotides. The following protocol describes the use of the overlap-extension PCR technique [13, 14] to create an insertion site in the cDNA of the target protein, which is required for the generation of the tension sensor and FRET control constructs (Fig. 2b, h, i) (*see* **Note 4**).

1. Design the following oligonucleotides (*see* Fig. 3):
   - Forward primer (F): matches the first 20–30 nucleotides (nts) of the cDNA sense strand (5′-end); the 5′-end encodes for a suitable restriction site and a Kozak-ATG sequence.
   - Reverse primer (R): matches the first 20–30 nts of the cDNA antisense strand (5′-end); the 5′-end encodes for a suitable restriction site and a stop codon.
   - Forward insertion primer (iF): the 5′-end of this primer encodes for the sense strand of the insertion fragment; the remaining nucleotides match 25–35 nts of the cDNA sense strand downstream of the insertion site.
   - Reverse insertion primer (iR): the 5′-end of this primer encodes for the antisense strand of the insertion fragment; the remaining nucleotides match 25–35 nts of the cDNA antisense strand downstream of the insertion site.

2. Prepare the following reaction mix to generate products A and B (*see* Fig. 3a): 40 μl ultrapure water, 5 μl 10× PfuUltra II Rxn buffer, 1 μl primer F (or iF), 1 μl primer iR (or R), 1 μl dNTPs, 1 μl cDNA template (5–30 ng/μl), 1 μl PfuUltra II Fusion HS DNA Polymerase.

3. Run touch-down PCR using the program below:
   - Step 1: 95 °C for 180 s.
   - Step 2: 95 °C for 20 s.
   - Step 3: 68 °C for 20 s (−1 °C every cycle).

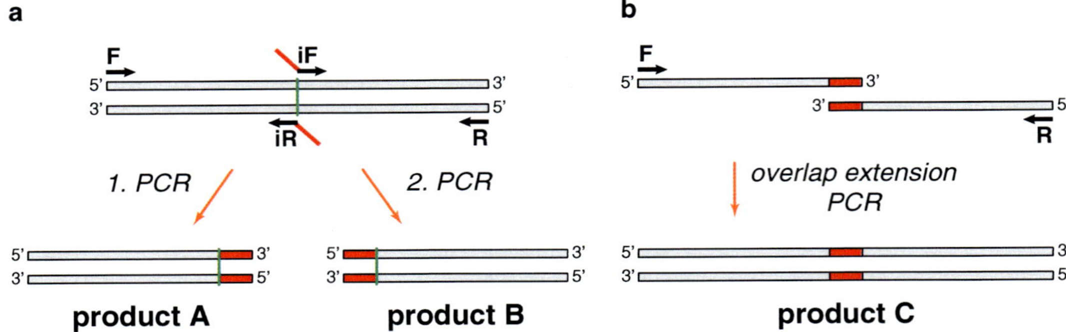

Fig. 3 Schematic view of the overlap-extension PCR. *Gray bars* indicate sense and antisense strand of the template cDNA, *arrows* indicate forward (F) and reverse (R) primer. The *red line* of insertion primers (iF and iR) indicates the nonbinding insertion sequence, the *green line* marks the insertion site. (**a**) In the first step, two modified cDNA products are generated (product A and product B). The modified sequence is indicated by the *red bar*. (**b**) The overlap-extension PCR uses the cDNA products from (**a**) as templates and F and R as primers. The resulting cDNA (product C) contains the modified sequence indicated in *red*

- Step 4: 72 °C for 15 s/1 kb of product (repeat **steps 2–4**; 8×).

- Step 5: 95 °C for 20 s.

- Step 6: 60 °C for 20 s.

- Step 7: 72 °C for 15 s/1 kb of product (repeat **steps 5–7**; 30×).

- Step 8: 72 °C for 180 s.

- Step 9: hold at 4 °C.

4. Run gel electrophoresis of PCR products using a suitable DNA ladder to confirm the size. Isolate correct bands and determine DNA concentration. Dilute the products to a concentration of at least 50 ng/µl (*see* **Note 5**).

5. Prepare the following reaction mix to generate product C (*see* Fig. 3b): 39 µl ultrapure water, 5 µl 10× PfuUltra II Rxn buffer, 1 µl primer F, 1 µl primer R, 1 µl dNTPs, 1 µl product A, 1 µl product B, 1 µl PfuUltra II Fusion HS DNA Polymerase.

6. Run touch-down PCR using the program described in **step 3**. Adjust elongation time for the expected length of product C.

7. Confirm the size of product C using DNA gel electrophoresis and isolate the DNA fragment.

8. Digest product C with the appropriate restriction enzymes and subclone into pBluescript II plasmid; confirm correct sequence by DNA sequencing.

9. Insert tension sensor module (Fig. 2b) or individual fluorophores (Fig. 2h, i) by standard cloning techniques and transfer the final construct into a suitable expression vector (*see* **Note 6**).

3.2 Expression of the Biosensor Constructs in Living Cells

Depending on the experimental requirements biosensor and control constructs can be either transiently or stably expressed. For transient expression, a variety of transfection methods have been described with transfection efficiencies strongly depending on the cell line of choice. Lipofectamine 2000 usually provides good results with rather low toxicity in most cell types. For stable expression replication-deficient retroviruses produced by transiently transfected Gryphon cells are used. The following protocols are optimized for fibroblastoid murine cell lines.

3.2.1 Transient Expression

The following procedure describes transfection in 6-well plates. Quantities refer to a single well.

1. Seed cells 16–24 h prior to transfection in 2 ml antibiotic-free transfection medium. Cells should be approximately 80 % confluent at the time of transfection.

2. Dilute 4 µg DNA in 250 µl Opti-MEM medium using a 1.5 ml reaction tube (*see* **Note 7**).

3. Gently mix Lipofectamine 2000 and dilute 10 µl in 250 µl Opti-MEM medium using a separate 1.5 ml reaction tube. Incubate for 5 min at room temperature.

4. Combine DNA and Lipofectamine 2000 solutions, mix gently and incubate for 20 min at room temperature.

5. Add the transfection mix dropwise to the cells and move the plate gently back and forth to ensure a homogenous distribution (avoid circular motion). Incubate cells at 37 °C (5 % CO_2).

6. Replace medium with growth medium after 6 h. Cells can be analyzed 24–48 h after transfection.

3.2.2 Stable Expression

The following protocol uses Gryphon cells (*see* **Notes 8** and **9**) to produce replication-deficient retroviruses [15].

1. Seed approximately 3.3×10^6 Gryphon cells onto a 10 cm dish and incubate at 37 °C (5 % CO_2) overnight. It is recommended that the cells are approximately 80 % confluent at the time of transfection.

2. Plate target cells onto a 10 cm dish. Do not seed cells too dense as cell proliferation and nuclear breakdown are critical for stable integration into the host cell genome.

3. Prior to transfection, remove the old medium from Gryphon cells and add 7 ml fresh growth medium supplemented with 4 µl of 50 mM chloroquine. After the addition of the transfection cocktail (**step 6**), the final chloroquine concentration will be 25 µM (*see* **Note 10**).

4. Dilute 20 μg of DNA with ultrapure water to a total amount of 439 μl in a 2 ml reaction tube; add 61 μl of 2 M $CaCl_2$ and vortex.

5. While vortexing, add 500 μl of 2×HBS dropwise to the DNA/$CaCl_2$ solution.

6. Add the transfection cocktail dropwise to the cells and move the plate back and forth to ensure a homogenous distribution (avoid circular motion). Incubate cells at 37 °C (5 % CO_2).

7. Replace transfection medium with the regular growth medium after 5–6 h. Incubate cells at 37 °C (5 % CO_2) overnight.

8. The next morning, replace medium on Gryphon cells with 8 ml of fresh growth medium (*see* **Note 11**).

9. After 12 h collect retroviral supernatant and add 8 ml of fresh growth medium to Gryphon cells.

10. Add 8 μl polybrene to the collected supernatant. Sterile filter the solution to remove detached Gryphon cells and to avoid contamination.

11. Replace medium on target cells with the collected supernatant from **step 10**.

12. Repeat **steps 9–11** in the morning and in the evening of the following day. In total, retroviral supernatant is added three times to the target cells.

13. Replace the virus containing medium with normal growth medium the next morning. Protein expression can be expected within 24 h after the last infection. Positive cells may be selected by fluorescence activated cell sorting (FACS) or antibiotic selection depending on the used expression vector.

3.3 Biosensor Evaluation

3.3.1 Biochemical Evaluation

The following protocols are optimized for cytoskeletal proteins with molecular weights between 20 and 250 kDa. Protocols may have to be empirically optimized depending on the biochemical properties of protein of interest.

1. Seed cells in a 6-well plate and allow them to attach.

2. Remove medium and wash once with ice-cold PBS.

3. Lyse cells in 200 μl ice-cold lysis buffer for 10 min; use a cell scraper to detach cells.

4. Transfer lysate to a 1.5 ml reaction tube and centrifuge at 16,100×g for 10 min at 4 °C.

5. Transfer supernatant into a fresh 1.5 ml reaction tube (keep on ice), determine protein concentration, adjust protein levels if required, add loading buffer, and boil samples at 95 °C for 10 min.

6. Run SDS-polyacrylamide gel electrophoresis (PAGE) and blot gel onto a PVDF membrane. Block the membrane for 1 h at

room temperature in blocking solution and incubate overnight at 4 °C with the anti-GFP antibody (diluted 1:1,000 in blocking solution) (*see* **Notes 1** and **13**).

7. Wash membrane (3× for 5 min in TBS-T) and incubate for 1.5 h at room temperature with HRP-coupled secondary antibody (diluted 1:10,000 in blocking solution) (*see* **Note 13**).

8. Wash membrane (3× for 5 min in TBS-T) and develop the signal using a chemiluminescence-based detection kit and record it on film or digitally. Quantify protein expression levels by densitometric analysis.

3.3.2 Immuno-histochemical Evaluation

The following protocol describes immunostaining of adherent cells in 24-well plates. Quantities refer to a single well.

1. Clean coverslips by shaking in 100 % ethanol containing 0.1 % NaOH for 30 min, then wash twice for 30 min in ultrapure water. Coverslips can be stored in 100 % ethanol at room temperature. Flame coverslips before placing into a 24-well plate.

2. Coat coverslip with 200 μl FN solution for 30 min at room temperature or overnight at 4 °C.

3. Seed cells onto coverslips and allow them to attach. Most cell lines will attach within 1–2 h (*see* **Note 12**).

4. Remove medium and wash cells with PBS. Fix in ice-cold 4 % PFA for 10 min (*see* **Note 2**).

5. Wash cells 3× for 5 min in PBS, block for 1 h and incubate with primary antibody (diluted 1:400 in blocking solution) for 1 h at room temperature (*see* **Note 13**).

6. Wash cells 3× for 5 min in PBS and incubate with fluorescently labeled secondary antibody (diluted 1:400 in blocking solution) for 1 h at room temperature (*see* **Notes 13** and **14**).

7. Wash cells 3× for 5 min in PBS.

8. Mount coverslip (cells downward) on a microscope slide using 5 μl mounting medium and let them dry overnight at room temperature. Keep slides at 4 °C in the dark to avoid photobleaching (*see* **Note 15**).

9. Evaluate immunostaining at the microscope using an oil-immersion objective. Especially the localization of tension sensor construct (Fig. 2b) as compared to the wt-control (Fig. 2c) should be carefully analyzed and, if possible, quantified.

3.4 Live Cell Microscopy and Image Acquisition

Glass-bottom cell culture dishes are commercially available but can be cost-effectively prepared in the laboratory (*see* **Note 16**).

1. Coat glass coverslip with 400 μl of FN solution for 30 min at room temperature or overnight at 4 °C. After washing once with PBS add 2 ml PBS and sterilize the dish by 20 min exposure to UV light.

2. Remove PBS, seed $1-5 \times 10^5$ cells in 2 ml of live-cell-imaging medium and allow them to adhere. Change medium to remove nonadherent cells before microscopy.

3. Mount glass-bottom dishes onto the microscope stage. To obtain low background signal it is recommended to use a water objective (1.2 NA). Allow the system to equilibrate for a couple of minutes to avoid focus drift during the experiment.

4. Adjust the microscope settings using the detection system of the scanning microscope. Individual parameters should be optimized for each biosensor. Typically, a scanning rate of 400 Hz, a resolution of 512×512 pixels and a digital zoom factor of 2 can be used. The laser power of the white light laser should not exceed 50–75 % to minimize photobleaching of the sample. If the sample is too dim, the pinhole (usually set to 1 Airy Unit) should be opened (*see* **Note 17**).

5. Using the eyepiece of the microscope select cells with appropriate expression levels (*see* **Note 18**) and adjust the focus. Avoid extended exposure of cells with excitation light to reduce photobleaching.

6. Acquire an image using the scanning microscope detection system.

7. Switch to FLIM detection and acquire a single image or a series of images depending on the brightness of the sample using the Imspector software. To achieve sufficient photon counts 5–15 consecutive images can be recorded and averaged to boost statistical power; this may require an exposure time of about 5–45 s.

8. Save the data in the .tif format for subsequent image processing using the autosave function of the Imspector software.

9. Repeat **steps 5–8** until a sufficient number of images are acquired (*see* **Note 19**).

3.5 Data Analysis and Evaluation

Data can be immediately analyzed using the Imspector software or processed later using custom-written software. It is recommended to check the data structure directly at the microscope using the Imspector software; for a detailed analysis images may be exported and separately analyzed in MATLAB. Note that fluorescence lifetimes can be calculated in different ways (*see* **Notes 20** and **21**).

1. Export data from the Imspector software as a series of 153 .tif images and import them into MATLAB. Use the LOCI package to import and process metadata.

2. First estimation of the fluorescence lifetimes is achieved by analyzing the arrival time of all photons regardless of their spatial position (Fig. 4a, b). Using a custom-written MATLAB routine, fit a mono-exponential decay curve to the histogram

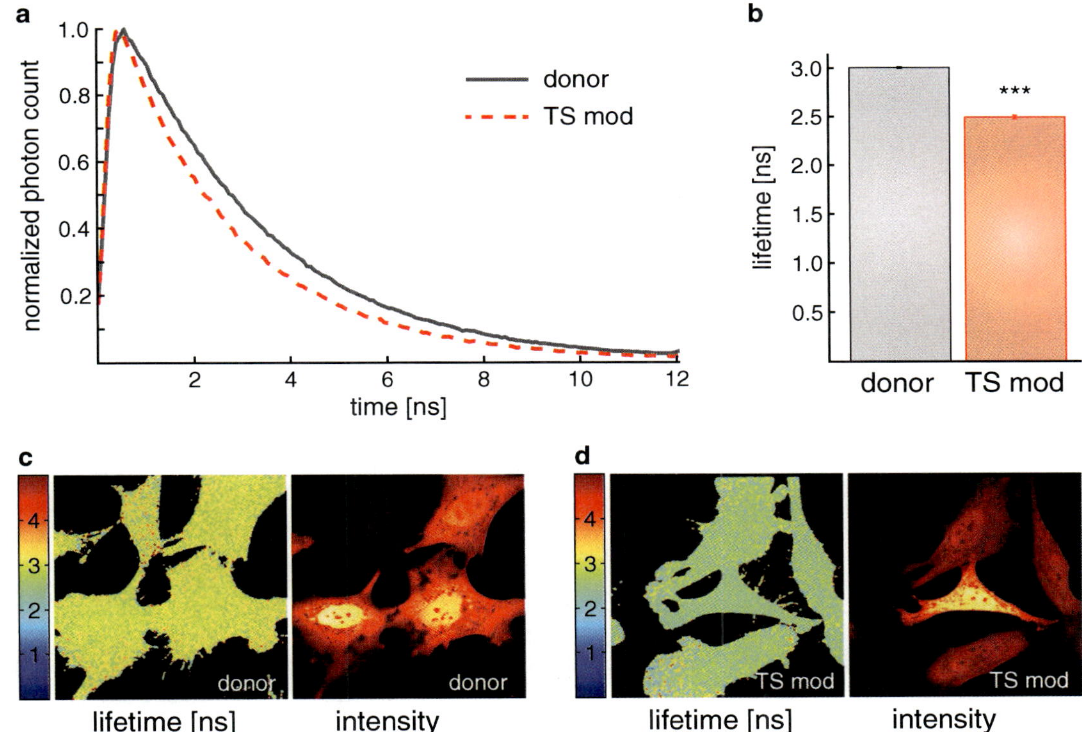

Fig. 4 Quantitative analysis of FLIM data. (**a**) Distribution of photon arrival times from cells expressing only the donor fluorophore or the tension sensor module (TS mod). (**b**) Photon arrival times from cells ($n=10$) expressing only the donor fluorophore or the tension sensor module (TS mod) were used to calculated the average fluorescence lifetime. Error bars indicate the standard error of the mean (SEM); *stars* indicate significance of $p<0.001$ in the Student's t-test. (**c**) Representative lifetime and intensity image of cells expressing only a donor fluorophore. (**d**) Representative lifetime and intensity image of cells expressing a tension sensor module (TS mod). Note that calculated lifetimes are independent of the expression level as expected

of photon arrival times by a nonlinear least square fit using the Levenberg-Marquardt algorithm.

3. Use fluorescent lifetimes of cells expressing only the donor fluorophore (τ_D) (Fig. 2c, f) (*see* **Note 22**) or the tension sensor controls (τ_{DA}) (Fig. 2d, e) (*see* **Note 23**) to determine the FRET efficiency (E) of the tension sensor under zero-force conditions using: $E = 1 - (\tau_{DA}/\tau_D)$.

4. Use cells co-expressing intermolecular FRET constructs (Fig. 2h, i) to estimate intermolecular FRET (*see* **Note 24**).

5. Calculate fluorescence lifetimes in cells expressing the wt-tension sensor (Fig. 2b) and compare with cells expressing the tension sensor controls (Fig. 2d, e). An increase in fluorescence lifetime (and a decrease in FRET efficiency) in the wt-tension sensor-expressing cells is indicative of force acting upon the protein of interest.

4   Notes

1. The recommended antibody (ab290, Abcam) recognizes GFPs and yellow fluorescent proteins (YFPs), but fails to bind cyan fluorescent proteins (CFPs) or red fluorescent proteins (RFPs).

2. The PFA solution should be freshly prepared and kept on ice until use. Alternatively, aliquots can be stored at −20 °C.

3. The band pass filter should efficiently block emission light of the acceptor fluorophore. Filters can be tested using cells expressing only the acceptor fluorophore.

4. The insertion site ideally contains two restriction sites that can be used to introduce the tension sensor module (Fig. 2b) or single fluorophores (Fig. 2h, i) by standard cloning techniques. Restriction sites should be separated by at least three nucleotides to allow double digest. The insertion must not cause a frame-shift and stop codons have to be removed.

5. A tenfold dilution of product A and B usually yields good results of the final PCR.

6. For transient transfection pcDNA3.1 and for stable transduction pLPCX expression vectors can be used.

7. Repeated freeze–thaw cycles can reduce DNA quality. In the case of unsatisfactory transfection efficiencies prepare fresh plasmid DNA. For especially large expression constructs increase the amount of DNA.

8. Gryphon cells can be cultured in a broad range of media. Ideally, they are cultured in medium optimized for target cells. Due to high proliferation rate and formation of cell–cell contacts it is recommended to routinely split Gryphon cells before they reach confluency. All liquids have to be added slowly to avoid detachment of cells.

9. Retrovirus-producing cell lines such as Gryphon cells can have an ecotropic or an amphotropic host range. To infect human cells amphotropic Gryphon cells have to be used. Appropriate safety measures must be taken when working with amphotropic virions.

10. Extended chloroquine treatment is toxic and will reduce the retroviral titer. Therefore, the chloroquine treatment should not exceed 6–12 h [15].

11. Transfection efficiency should be 80–100 % and can be confirmed using a fluorescence microscope about 24 h after transfection.

12. Some cell lines will adhere to plastic and glass surfaces without subsequent surface coating. Other cell lines may not adhere to FN because they lack the required cell surface receptors; these cells should be seeded on other extracellular matrices such as vitronectin, collagen, or laminin.

13. If necessary, the optimal concentration of the primary and secondary antibodies should be determined empirically.

14. The excitation and emission properties of labeled secondary antibodies should not interfere with the donor and acceptor fluorophores.

15. Immunostainings can be stored for several weeks. It is recommended, however, to analyze stainings as soon as possible since the fluorescence signal will decay over time.

16. To prepare glass-bottom cell culture dishes, a 15 mm hole is drilled into the bottom of a 35 mm Petri dish. After a thin film of RV1 silicone glue has been applied at the edge of the hole (bottom side), a cleaned coverslip is pressed onto the glue with an evenly distributed force. Dishes should dry overnight at room temperature before use.

17. Microscope and laser settings should not be changed during the course of the experiment. Especially images of cells expressing the tension sensor (Fig. 2b) and control constructs (Fig. 2c–g) should be recorded using identical settings.

18. Avoid cells overexpressing the constructs, as they do not reflect the physiological condition.

19. 10–30 images should be acquired for each construct to achieve sufficient statistical power.

20. Alternative methods to evaluate fluorescence lifetime include maximum likelihood estimators (MLE) or the phasor approach [16].

21. FLIM data analysis can be restricted to a region of interest, for instance, by thresholding the signal. This is of special interest for biological processes confined to a subcellular location.

22. The fluorescence lifetimes of currently used donor fluorophores such as CFPs, GFPs, or YFPs are typically in the range of 2–3 ns.

23. In the presence of the acceptor fluorophore the fluorescence lifetime of the donor should decrease by 20–30 %.

24. Intermolecular FRET should be negligible. Strongly overexpressing cells often display intermolecular FRET and should be excluded from the analysis.

References

1. Hoffman BD, Grashoff C, Schwartz MA (2011) Dynamic molecular processes mediate cellular mechanotransduction. Nature 475:316–323

2. Bao G, Suresh S (2003) Cell and molecular mechanics of biological materials. Nat Mater 2:715–725

3. Wang JH, Li B (2009) Application of cell traction force microscopy for cell biology research. Methods Mol Biol 586:301–313

4. Rowat AC (2009) Physical properties of the nucleus studied by micropipette aspiration. Methods Mol Biol 464:3–12

5. Mackay JL, Kumar S (2013) Measuring the elastic properties of living cells with atomic force microscopy indentation. Methods Mol Biol 931:313–329

6. Kasza KE, Vader D, Koster S et al (2011) Magnetic twisting cytometry. Cold Spring Harb Protoc 2011:pdp prot5599

7. Grashoff C, Hoffman BD, Brenner MD et al (2010) Measuring mechanical tension across vinculin reveals regulation of focal adhesion dynamics. Nature 466:263–266

8. Doyle AD, Yamada KM (2010) Cell biology: Sensing tension. Nature 466:192–193

9. Jares-Erijman EA, Jovin TM (2006) Imaging molecular interactions in living cells by FRET microscopy. Curr Opin Chem Biol 10:409–416

10. Jares-Erijman EA, Jovin TM (2003) FRET imaging. Nat Biotechnol 21:1387–1395

11. Morton PE, Parsons M (2011) Measuring FRET using time-resolved FLIM. Methods Mol Biol 769:403–413

12. Becker W (2012) Fluorescence lifetime imaging – techniques and applications. J Microsc 247:119–136

13. Nelson MD, Fitch DH (2011) Overlap extension PCR: an efficient method for transgene construction. Methods Mol Biol 772:459–470

14. Heckman KL, Pease LR (2007) Gene splicing and mutagenesis by PCR-driven overlap extension. Nat Protoc 2:924–932

15. Swift S, Lorens J, Achacoso P, et al (2001) Rapid production of retroviruses for efficient gene delivery to mammalian cells using 293T cell-based systems. In: John E. Coligan et al. (eds.) Current protocols in immunology. Chapter 10, Unit 10 17C

16. Digman MA, Caiolfa VR, Zamai M et al (2008) The phasor approach to fluorescence lifetime imaging analysis. Biophys J 94:L14–16

Chapter 16

Proteomic Analysis of the Left Ventricle Post-myocardial Infarction to Identify In Vivo Candidate Matrix Metalloproteinase Substrates

Andriy Yabluchanskiy, Yaojun Li, Lisandra E. de Castro Brás, Kevin Hakala, Susan T. Weintraub, and Merry L. Lindsey

Abstract

Left ventricular remodeling post-myocardial infarction (MI) involves a multitude of mechanisms that regulate the repair response. Matrix metalloproteinases (MMPs) are a major family of proteolytic enzymes that coordinate extracellular matrix turnover. MMP-7 or MMP-9 deletion attenuate adverse remodeling post-MI, but the mechanisms have not been fully clarified. Both MMP-7 and MMP-9 have a large number of known in vitro substrates, but in vivo substrates for these two MMPs in the myocardial infarction setting are incompletely identified. Advances in proteomic techniques have enabled comprehensive profiling of protein expression in cells and tissue. In this chapter, we describe a protocol for the proteomic analysis of in vivo candidate MMP substrates in the post-MI left ventricle using two-dimensional electrophoresis, liquid chromatography coupled with tandem mass spectrometry, and immunoblotting.

Key words Proteomics, Cardiac remodeling, Mice, Extracellular matrix, Matrix metalloproteinase, Myocardial infarction, Substrates

1 Introduction

A disruption of blood supply in a coronary artery can lead to irreversible ischemia and the development of myocardial infarction (MI). Following MI, the left ventricle (LV) undergoes extensive remodeling, which results in reorganization and accumulation of new extracellular matrix (ECM) proteins that attempt to sustain cardiac integrity and function [1–3]. Matrix metalloproteinases (MMPs) are a family of zinc-dependent proteolytic enzymes capable of processing ECM proteins [4]. As such, MMPs orchestrate ECM turnover post-MI [5, 6]. The absence of several specific

Andriy Yabluchanskiy and Yaojun Li have contributed equally to this chapter.

Troy A. Baudino (ed.), *Cell-Cell Interactions: Methods and Protocols*, Methods in Molecular Biology, vol. 1066, DOI 10.1007/978-1-62703-604-7_16, © Springer Science+Business Media New York 2013

185

MMP family members, including MMP-7 and MMP-9, has positive effects on LV remodeling post-MI [7, 8]. However, the biological mechanisms have not been fully clarified. In vitro, MMP-7 is capable of cleaving a broad range of ECM proteins, including collagen IV, fibronectin, laminin, and tenascin-C [9–12], as well as non-ECM proteins, including tissue factor pathway inhibtor-2, E-cadherin, and MMP-2 [13–15]. Similar to MMP-7, MMP-9 in vitro cleaves many ECM proteins, including gelatin, collagen IV, and collagen V [16–18], and cleaves non-ECM proteins, including pro-TGFβ, pro-TNFα, and other MMPs [19–21]. However, in vivo substrates for MMP-7 and MMP-9 post-MI have not been fully identified.

In this book chapter, we describe a protocol that uses two-dimensional electrophoresis (2-DE), along with high performance liquid chromatography coupled with electrospray tandem mass spectrometry (HPLC-ESI-MS/MS), to identify in vivo MMPs substrates in the infarcted myocardium [22, 23]. The discovery of new in vivo MMPs substrates would potentially identify novel therapeutic targets to reduce adverse LV remodeling.

2　Materials

2.1　Mouse Model of MI

1. C57BL/6 J WT, MMP-7 null, or MMP-9 null mice, 4–6 months of age were used. Both MMP-7 null mice and MMP-9 null mice were gifts from Dr. Lynn Matrisian from Vanderbilt University [24, 25].

2. Isoflurane.

3. Sterile polyamide monofilament 8-0 suture.

4. High-frequency ultrasound system (e.g., Vevo 770® High-Resolution Imaging System; VisualSonics, Toronto, Ontario, Canada).

5. Biofeedback surgery platform.

6. Anesthesia system.

7. Electrocardiogram recording system.

8. Buprenorphine.

2.2　Infarct Tissue Collection

1. Isoflurane.

2. 2,3,5-Triphenyltertrazolium chloride (TTC).

2.3　Protein Extraction

1. Protein extraction reagent type 4: 7 M urea, 2 M thiourea, 40 mM trizma base and 1 % C7BzO (Sigma-Aldrich, St. Louis, MO, USA).

2. Complete protease inhibitor cocktail (Roche, Madison, WI, USA).

3. Bradford assay reagent.

4. Homogenizer.

2.4 Two-Dimensional Electrophoresis

1. Protein extraction reagent type 4 (Sigma-Aldrich).

2. Complete protease inhibitor cocktail (Roche).

3. Tributylphosphine.

4. Iodoacetamide.

5. Immobilized pH gradient (IPG) strip pH 3–10 (Bio-Rad Laboratories, Hercules, CA, USA).

6. Mineral oil.

7. Criterion XT 4–12 % bis–tris precast gels (Bio-Rad Laboratories).

8. Kaleidoscope prestained molecular weight standard (Bio-Rad Laboratories).

9. XT MOPS running buffer (Bio-Rad Laboratories).

10. XT MES running buffer (Bio-Rad Laboratories).

11. EZ Blue gel staining reagent (Sigma-Aldrich).

12. Centrifugal vacuum concentrator.

13. Electrophoresis power supply.

14. Gel imaging system.

15. 2-DE gel spot detection/analysis software (e.g., Progenesis PG240; Nonlinear Dynamics, Durham, NC, USA).

2.5 Identification of Proteins in 2-DE Gel Spots by Tandem Mass Spectrometry

1. Sequencing-grade trypsin.

2. Gel spot excision tool (e.g., One Touch™ 1.5 mm 2-DE spot picker; Gel Company, San Francisco, CA, USA).

3. Rotator.

4. Centrifugal vacuum concentrator.

5. Electrospray tandem mass spectrometer (e.g., LTQ, Thermo Fisher Scientific, Hudson, NH, USA).

6. Nanoflow HPLC system (e.g., NanoLC micro-HPLC; Eksigent/AB Sciex, Framingham, MA, USA).

7. Capillary HPLC column [e.g., self-packed PicoFrit (New Objective, Woburn, MA, USA; 75 μm i.d.) packed to 10 cm with C_{18} adsorbent (Grace Vydac, Hesperia, CA, USA; 5 μm, 300 Å)].

8. Mobile phases: A, 0.5 % acetic acid/0.005 % trifluoroacetic acid (TFA); B, 90 % acetonitrile/0.5 % acetic acid /0.005 % TFA.

9. Database search software (e.g., Mascot; Matrix Science, Boston, MA, USA).

10. Post-processing software (e.g., Scaffold; Proteome Software, Portland, OR, USA).

2.6 Immunoblotting Analysis of In Vivo Candidate MMP Substrates

1. 10× Tris/glycine buffer.
2. Criterion XT bis–tris gel, 18- or 26-well, 4–12 % (Bio-Rad Laboratories).
3. XT MOPS running buffer (Bio-Rad Laboratories).
4. XT MES running buffer (Bio-Rad Laboratories).
5. MemCode reversible protein stain kit (Pierce, Rockford, IL, USA).
6. Blotting grade blocker (e.g., nonfat dry milk).
7. Primary antibodies of the MMP substrates (e.g., anti-fibronectin).
8. Enhanced chemiluminescent (ECL) substrate for horseradish peroxidase (HRP).
9. Blotting paper.
10. Nitrocellulose membrane.
11. Criterion blotter (Bio-Rad Laboratories).
12. Imaging software.

2.7 In Vitro and Ex Vivo MMP Cleavage Assay

1. Recombinant human MMP-7 (Calibiochem, EMD Millipore, Darmstadt, Germany).
2. Recombinant human MMP-9 (Calibiochem).
3. Recombinant human proteins (e.g., fibronectin, R&D Systems, Minneapolis, MN, USA).
4. Zymogram developing buffer (Invitrogen, Grand Island, NY, USA).
5. XT sample buffer (Bio-Rad Laboratories).
6. Recombinant human MMP-7 (Calibiochem).
7. Recombinant human MMP-9 (Calibiochem).
8. XT sample buffer (Bio-Rad Laboratories).
9. Criterion Bis–Tris gel, 26-well, 4–12 % (Bio-Rad Laboratories).
10. XT MOPS running buffer (Bio-Rad Laboratories).
11. XT MES running buffer (Bio-Rad Laboratories).
12. 10× Tris/glycine buffer.
13. Blotting grade blocker.
14. Primary antibodies of the MMP substrates (e.g., anti-fibronectin).
15. ECL substrate for HRP.
16. Nitrocellulose membrane.
17. Filter paper.
18. Criterion blotter (Bio-Rad Laboratories).
19. Imaging software.

3 Methods

3.1 Mouse Model of Myocardial Infarction

1. Induce MI by coronary artery ligation on wild-type and null mice, as previously described [8]. In brief, coronary ligation is made by using a sterile polyamide monofilament 8-0 suture to ligate the left coronary artery and disrupt the blood supply.

2. Confirm MI by looking for visible LV blanching and ST segment elevation by electrocardiography.

3. Inject 0.05 mg/kg of buprenorphine intraperitoneally into the mouse. Place the mouse in a 37 °C incubator with room air supplemented with oxygen for recovery.

4. Monitor continuously until the mouse is alert and ambulatory, at which point the mouse is returned to its cage and monitored daily.

3.2 Infarct Tissue Collection

1. Mice that survive through day 7 undergo a terminal echocardiographic analysis as previously described [8] and are then sacrificed for subsequent analyses.

2. Anesthetize each mouse with 2 % isoflurane (*see* **Note 1**), flush the coronary vasculature with phosphate buffered saline (PBS), and excise the heart.

3. Separate the LV from the right ventricle and weigh each individually.

4. To measure infarct size, section the LV into three slices, stain with 1 % TTC for 5 min at 37 °C, and photograph the slices.

5. Separate the infarct tissue from the remote tissue, then weigh, snap-freeze in liquid nitrogen, and store at −80 °C in individual tubes for further analysis.

3.3 Protein Extraction

1. To prepare the Protein Extraction Reagent Type 4 (Reagent 4), add 15 ml of deionized water, swirl and heat to 30 °C for 30–60 min until dissolved. Use immediately or aliquot (1 ml/tube) and store at −20 °C.

2. To prepare the 10× protease inhibitor, dissolve one tablet of Complete Protease Inhibitor Cocktail in 1 ml of deionized water and vortex until fully dissolved, then aliquot (0.5 ml/tube) and store at −20 °C.

3. Record the wet weights of the LV samples and place each in a 5 ml polypropylene round-bottom tube.

4. Add 800 μl of lysis buffer (720 μl of Reagent 4 and 80 μl of 10× protease inhibitor) per 50 mg of tissue and homogenize.

5. Transfer to a 1.5 ml polypropylene microcentrifuge tube and store at room temperature until all samples are homogenized.

6. Heat samples at 30 °C for 15 min and snap freeze.

7. Thaw samples at 30 °C for 30 min before determining the protein concentration of each sample by the Bradford assay following the manufacturer's instructions (*see* **Note 2**).

3.4 Two-Dimensional Gel Electrophoresis

1. Mix 500 μg of each protein sample (~80 μl) with protease inhibitor cocktail (1× final), 1:40 (v/v) tributylphosphine solution, and water to make a total volume of 200 μl, and reduce proteins at room temperature for 1 h.

2. Add 1:10 (v/v) 10× protease inhibitor and 1:33 (v/v) iodoacetamide solution (93.3 mg/ml) and alkylate in the dark at room temperature for 1 h (*see* **Note 3**).

3. Spin the sample at low speed (425 × *g*) for 5 min to pellet debris.

4. Transfer each supernatant into a new tube and discard the tubes with the pellets (*see* **Note 4**).

5. Add acetone to a final concentration of 80 % and place at room temperature for 30 min.

6. Centrifuge samples at 20,000 × *g* for 10 min and discard the supernatant.

7. Use a centrifugal vacuum concentrator to remove residual acetone for 5 min (or air-dry for 5–10 min) (*see* **Note 5**).

8. Resuspend samples in 200 μl of Reagent 4 and 1× protease inhibitor, vortex, and incubate at 30 °C for 30 min (*see* **Note 6**).

9. Add protein sample to the rehydration tray, then place an 11-cm IPG strip (pH 3–10) on top of the protein solution with the gel side facing down, and rehydrate overnight at room temperature (*see* **Note 7**).

10. On the next day, add mineral oil to wells on the isoelectric focusing (IEF) tray before inserting the strips.

11. Wet wicks (2/tray) with water and place at each end of the tray.

12. Place the IPG strip in the cell with gel side facing down (*see* **Note 8**).

13. Cover each IPG strip with mineral oil (*see* **Note 9**) and start the IEF voltage program. For an 11-cm IPG strip and an ElectrophoretIQ 2000, we use the following: 100 V, 1 h; 200 V, 1 h; 400 V, 2 h; 1,000 V, 2 h; 5,000 V, 12 h; 1,000 V, hold. The current is set to be constant at 2 mA.

14. After IEF is complete, blot the extra oil from the edges of the IPG strip using filter paper.

15. Put each strip in an equilibration tray slot, gel side down, and incubate in 5 ml of SDS equilibrating buffer (6 M urea, 0.05 M Tris-acetate, pH 7, 2 % SDS, and 0.0067 % bromophenol blue) with shaking for 10 min. Repeat (*see* **Note 10**).

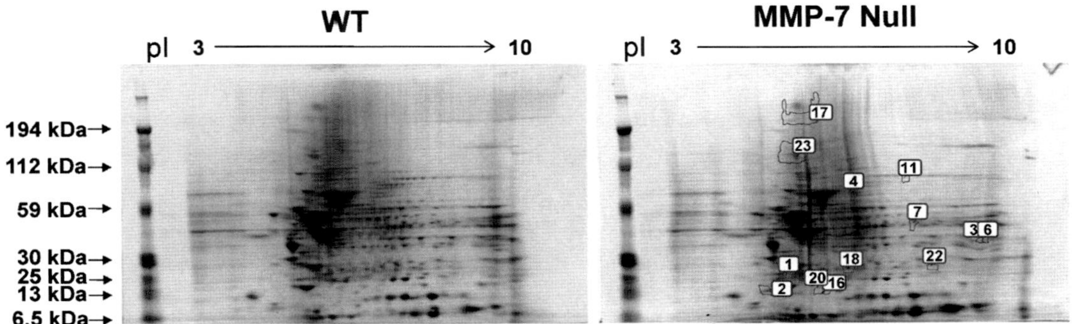

Fig. 1 Representative 2-DE gels of protein extracts from the infarct area of WT mice (*left*) and MMP-7 null mice (*right*). *Spot numbers* are indicated in the MMP-7 null gel. Reprinted with permission from [23]. Copyright 2010 American Chemical Society

16. Place an 11-cm precast gel (Criterion XT 4–12 % Bis–Tris) in an electrophoresis cell and add 1× XT MOPS running buffer up to the fill line.

17. Place the strip on the top of the precast gel and load 10 µl of Kaleidoscope Prestained Standard into the side well of the gel.

18. Run at 200 V (25–50 mA/gel) until the bromophenol blue reaches the bottom of the gel.

19. After the second dimension of electrophoresis, rinse each gel three times with deionized water for 5 min.

20. Fix each gel in 50 % methanol/10 % acetic acid at room temperature for 15 min.

21. Rinse each gel with deionized water for 15 min.

22. Add EZ Blue staining reagent to cover the gel and incubate for at least 45 min or overnight with shaking.

23. Destain each gel with deionized water until the background is clear, approximately 2 h.

24. Scan each gel and save images in the format compatible with the imaging software, using the exactly same parameters of the scanner and the exactly same parameters of the imaging software for each gel.

25. Analyze the 2-DE gel images with image analysis software (Fig. 1, *see* **Notes 11** and **12**).

3.5 Identification of Proteins in 2-DE Gel Spots by Mass Spectrometry

1. Excise gel spots with significant differences in intensity between the two groups either manually (e.g., using a One Touch 1.5-mm 2-DE Spot Picker) or with a robotic system.

2. Destain and wash each gel spot with 200 µl of 40 mM NH_4HCO_3/50 % acetonitrile in a 0.5 ml polypropylene microcentrifuge tube, and then slowly invert on a rotator for 20 min.

3. Remove the liquid and repeat the destaining and washing step once again (*see* **Note 13**).

4. Remove the liquid, quickly spin down at $1,000 \times g$, and remove the residual liquid.

5. Add 100 μl of acetonitrile to each tube and shake up and down until the gel spot turns white.

6. Dry each gel spot in a centrifugal vacuum concentrator for 5–10 min.

7. Swell the dried gel spots in 5–20 μl of 20 ng/μl trypsin in 40 mM NH_4HCO_3/20 % acetonitrile on ice and incubate for 40 min.

8. Add 10 μl of 40 mM NH_4HCO_3/20 % acetonitrile and incubate at 37 °C for 4 h.

9. Add 0.1 % TFA to the digests to a final volume of 200 μl and incubate at 37 °C for up to 1 h.

10. Spin down at $14,000 \times g$ for 5 min and transfer each supernatant to an autosampler vial and dry in a centrifugal vacuum concentrator.

11. Add 200 μl of 0.2 % TFA/50 % acetonitrile to each gel spot and incubate at 37 °C for up to 1 h as a second extraction.

12. Transfer each supernatant to the autosampler vial used for **step 10** and dry in a centrifugal vacuum concentrator.

13. Add 100 μl of 100 % acetonitrile to the gel spots and dehydrate them for the third extraction.

14. Transfer each supernatant to the autosampler vial used for **step 10** and dry in a centrifugal vacuum concentrator (*see* **Note 14**).

15. Resuspend the peptides in 12 μl of 0.5 % TFA just before HPLC-ESI-MS/MS analysis.

16. Analyze the digests by HPLC-ESI-MS/MS and separate the digests by reversed-phase capillary HPLC using a gradient of 2–42 % B in 30 min at a flow rate, 0.4 μl/min. (The method described here is for analysis using an Eksigent NanoLC micro-HPLC connected on-line to a Thermo Fisher LTQ linear ion trap mass spectrometer fitted with a New Objective PicoView 550 nanospray interface.)

17. Use a data-dependent scan strategy to collect precursor and tandem mass spectra. The settings used on a Thermo Fisher LTQ mass spectrometer are as follows: ESI voltage, 2.9 kV; isolation window for MS/MS, 3; relative collision energy, 35 %; scan strategy, survey scan followed by acquisition of data dependent collision-induced dissociation spectra of the seven most intense ions in the survey scan above a threshold of 3,000 intensity units.

18. (The parameters listed in **steps 18–20** are for database searching with Mascot.)

19. Process the mass spectrometry raw data file and generate a peak list. The program extractmsn.exe (Thermo Fisher Scientific) can be used with default settings for LTQ data.

20. Use a database search program such as Mascot to search the processed data file against a protein database, using either a complete database or a species-specific subset. The following parameters are appropriate for Mascot: proteolytic enzyme, trypsin; allowed missed cleavages, two; precursor mass tolerance, ±1.5 Da; fragment ion mass tolerance, ±0.8 Da; variable modifications, methionine oxidation and cysteine carbamidomethylation.

21. A post-processing program like Scaffold can be used to search the processed data file with X!Tandem and for cross-correlation with the Mascot results and determination of the probabilities of the peptide assignments and protein identifications.

3.6 Validation Step: Immunoblotting Analysis of In Vivo Candidate MMP Substrates

1. Place the samples at 30 °C for at least 30 min (*see* **Note 15**). Vortex and spin briefly at low speed to bring down any debris.

2. Prepare a positive control (*see* **Note 16**).

3. For each sample, load 10 μg of total protein in one lane of a 26-well 4–12 % Criterion Bis–Tris gel and run the gels in XT MES buffer or XT MOPS buffer at 200 V for 40 min (*see* **Note 17**).

4. After electrophoresis, remove the gels from the cassettes and rinse gently in deionized water.

5. Soak filter paper, sponges, and membrane in cold transfer buffer.

6. Place the layers in the following order, starting on the black side of the cassette: sponge, filter paper, gel, membrane, filter paper, sponge (*see* **Note 18**).

7. Transfer proteins at 65 V for 1 h (*see* **Note 19**).

8. After transfer, place the membrane in a clean container.

9. Stain the membrane with MemCode Reversible Protein Stain Kit as per the manufacturer's instructions.

10. Scan and analyze the membrane with imaging software and record the densitometry value of total proteins in each lane for normalizing the densitometry of each immunoblot (*see* **Notes 12** and **20**).

11. Destain the membrane with MemCode Reversible Protein Stain Kit according to the manufacturer's instructions.

12. Block nonspecific epitopes with 5 % nonfat dry milk in 1× PBS/0.1 % Tween-20 (blocking solution) at room temperature for 1 h with gentle shaking.

13. Wash the membrane with 1× PBS/0.1 % Tween-20 (PBST) for 10 min with gentle shaking, for a total of three wash steps.

14. Prepare the primary antibody by mixing the antibody at the concentration recommended by manufacturer (e.g., 1:2,000 for anti-fibronectin) in blocking solution. Incubate the membrane at room temperature for 2 h or at 4 °C overnight with gentle shaking.

15. Wash the membrane with 1× PBST for 10 min with gentle shaking and repeat, for a total of three washes.

16. Prepare the secondary antibody by mixing the antibody at the concentration recommended by the manufacturer in blocking solution. Incubate the membrane for 1 h at room temperature with gentle shaking.

17. Wash the membrane with 1× PBST for 10 min with gentle shaking and repeat, for a total of three washes.

18. Incubate the membrane with ECL substrate for 5 min and develop the film immediately in the dark room.

19. Scan and analyze the film with imaging software. Use the densitometry values to normalize the individual lanes for total protein (Fig. 2, *see* **Notes 12** and **20**).

3.7   In Vitro and Ex Vivo MMP-7 Cleavage Assay

1. For the in vitro MMP-7 cleavage assay, incubate 100 ng of the recombinant protein (e.g. recombinant fibronectin) with 10 pg, 100 pg, or 10 ng of recombinant human MMP-7 enzyme at 37 °C for 3 h in 1× zymogram developing buffer. Add 5 μl of SDS loading buffer to stop the reaction, followed by the **steps 3** and **4** in this section for immunoblotting (*see* **Note 21**).

2. For the ex vivo MMP-7 cleavage assay, incubate 10 μg of protein extract from the infarct area of an MMP-7 null mouse with 10 pg, 100 pg, or 10 ng of recombinant human MMP-7 enzyme at 37 °C for 3 h in 1× zymogram developing buffer (Fig. 3, *see* **Notes 12**, **21**, and **22**). Add 5 μl of SDS loading buffer to stop the reaction, followed by **steps 3** and **4** for immunoblotting.

3. Load each reaction mixture in a lane of a 26-well 4–12 % Criterion Bis–Tris gel, and run the gels in XT MES buffer or XT MOPS buffer at 200 V for 40 min (*see* **Note 17**).

4. Immunoblot the proteins of interest as described above in Subheading 3.6 to evaluate the action of exogenous MMP-7 enzyme on degradation of the substrate, comparing extracts from the control mouse to those from the MMP-7 null mouse (Fig. 3, *see* **Notes 21** and **22**).

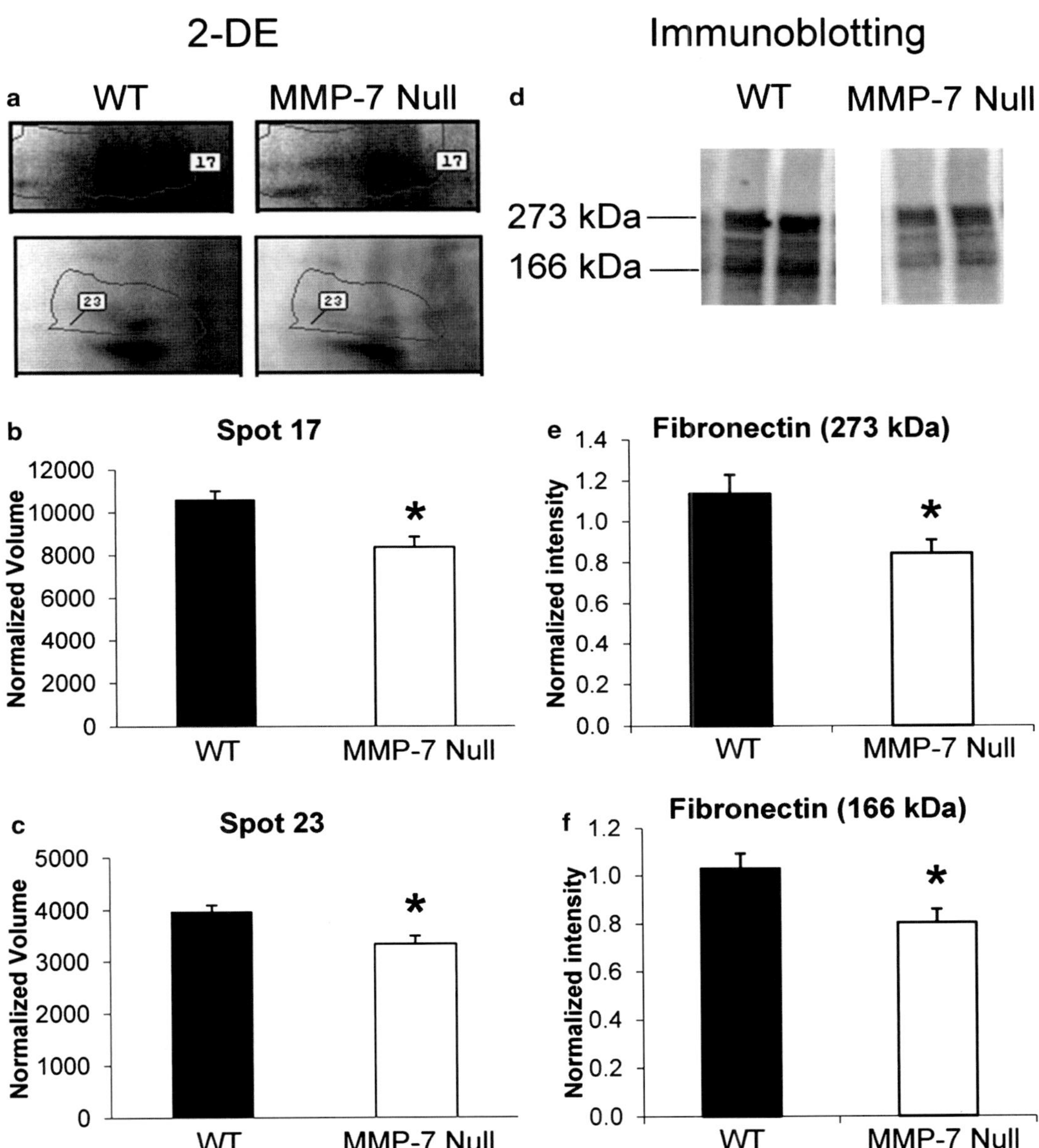

Fig. 2 The infarct area of MMP-7 null mice showed lower levels of fibronectin by 2-DE and immunoblotting. (**a**) Fibronectin was identified in spot 17 and spot 23 of the 2-DE gels. Both spot 17 (**b**) and spot 23 (**c**) showed significant lower intensity in the LV infarct area from MMP-7 null mice compared to WT. (**d**) Immunoblotting of fibronectin in the LV infarct area of WT and MMP-7 null mice were performed. The densitometry of the 273 kDa full-length (**e**) and the 166 kDa fragments (**f**) of fibronectin indicated both bands showed significantly lower intensity in the LV infarct area of MMP-7 null mice when compared with the WT ($p < 0.05$). Reprinted with permission from [23]. Copyright 2010 American Chemical Society

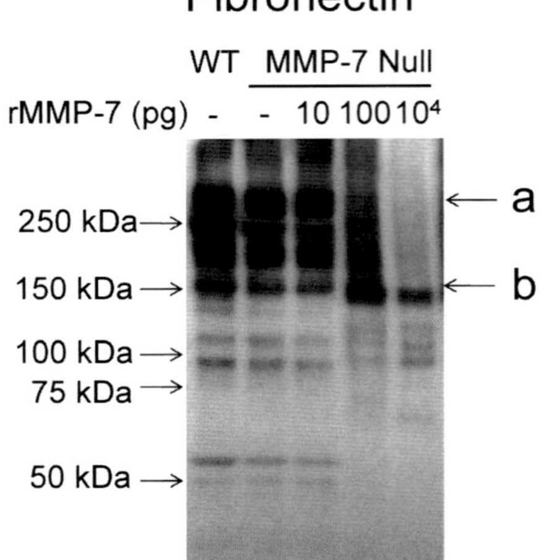

a – Full-length protein cleaved by MMP-7

b – Fragment generated by MMP-7 cleavage

Fig. 3 The addition of exogenous MMP-7 generated fibronectin fragments in protein extract from the infarct area of an MMP-7 null mouse. (**a**) Protein extract from the LV infarct area of an MMP-7 null mouse treated with 100 pg or 10 ng of recombinant MMP-7 showed reduced levels of full-length fibronectin (273 kDa) and increased levels of the 166 kDa fibronectin fragment compared to the untreated control. Reprinted with permission from [23]. Copyright 2010 American Chemical Society

4 Notes

1. Isoflurane is harmful to humans. Therefore, surgery should be performed in a well-ventilated room with a scavenging system, and the operator should turn off the anesthesia system when not in use.

2. Due to the high urea concentration in the extraction buffer, the protein extracts need to be diluted 1:40 with water prior to assaying for protein content to avoid interference with the colorimetric readings.

3. To make the iodoacetamide solution, add 600 μl of deionized water to dissolve the powder in a glass vial (56 mg/vial). Iodoacetamide should be prepared and used fresh. Discard any leftover solution.

4. This step must be completed prior to acetone precipitation.

5. The experimental process can be stopped here if needed and continued later.

6. Use 120 μl of Reagent 4 and 1× proteinase inhibitor for a 7 cm strip; 200 μl for an 11 cm strip; 300 μl for an 18 cm strip; and 440 μl for a 24 cm strip.

7. Make sure the rehydration tray is clean and dry, with no bubbles under the strips; and keep the tray in a horizontal position. Put the tray in a sealable plastic bag and place a wet paper towel in the bag to keep the strips moist (avoid drying out the strips). The strips can be frozen after rehydration and before equilibration.

8. Make sure the positive and negative poles are appropriately placed, with no bubbles at the end of the strip to prevent burning.

9. Make sure there is sufficient mineral oil to cover the strip and that the strip is well covered by the wicks.

10. Make sure that the equilibration buffer covers the strips, and remove the solution by aspiration.

11. When using the Progenesis PG240 software, image quality control can be performed to optimize image capture and image alignment, and the prefilter can be used to remove artifacts (e.g., gel areas with noise). The volume of each spot should be normalized to the total spot volume in the gel. Spots that exhibit significant differences between the two groups ($p < 0.05$ by Students t-test) are selected for protein identification by mass spectrometry.

12. All protein extracts should be analyzed individually and not pooled at any step. Analyze normalized spot volumes of 2-DE gels and normalized densitometries of the immunoblots between WT and null groups. A $p < 0.05$ is considered significant.

13. **Steps 1–3** in Subheading 3.5 may be repeated if necessary.

14. The dried peptides can be stored at room temperature in the dark.

15. Because of the high concentration of urea in Reagent 4, this step is needed to allow the samples to fully thaw in solution.

16. Spleen or tumor protein homogenates are often used as positive controls. If available, recombinant proteins are the most specific and best controls.

17. XT MES buffer is used for resolving proteins up to 100 kDa and XT MOPS buffer for proteins >100 kDa. Run times and voltages may differ, depending on the molecular weight of the protein of interests, the buffer, the electrophoresis apparatus, and the type of gel.

18. Make sure to roll out any bubbles while laying down the gel, the membrane, and the filter paper.

19. If the molecular weight of the target protein is more than 100 kDa, the transfer process will need to be modified to 65 V for 2 h or 40 V overnight.

20. Make sure that the blot is not overexposed and that the signal is in the dynamic range by the imaging software (e.g., Molecular Imaging Software version 4).

21. In vitro and ex vivo MMP-9 cleavage assays can be conducted using recombinant MMP-9, protein extract from LV infarct area of MMP-7 null mouse, and protein extract from LV infarct area of WT mouse.

22. Use a protein extract from the infarct area of an MMP-7 null mouse without MMP-7 treatment as the negative control, and use protein extract from the infarct area of a WT mouse without MMP-7 treatment as the positive control (Fig. 3).

Acknowledgments

We acknowledge support from NIH/NHLBI HHSN 268201000036C (N01-HV-00244) for the San Antonio Cardiovascular Proteomics Center and R01 HL075360, the Max and Minnie Tomerlin Voelcker Fund, and the Veteran's Administration (Merit) to M.L.L. Mass spectrometry analyses were conducted in the UTHSCSA Institutional Mass Spectrometry Laboratory.

References

1. Dixon JA, Spinale FG (2011) Myocardial remodeling: cellular and extracellular events and targets. Annu Rev Physiol 73:47–68

2. Lindsey ML, Weintraub ST, Lange RA (2012) Using extracellular matrix proteomics to understand left ventricular remodeling. Circ Cardiovasc Genet 5:o1–7

3. Gajarsa JJ, Kloner RA (2011) Left ventricular remodeling in the post-infarction heart: a review of cellular, molecular mechanisms, and therapeutic modalities. Heart Fail Rev 16:13–21

4. Page-McCaw A, Ewald AJ, Werb Z (2007) Matrix metalloproteinases and the regulation of tissue remodelling. Nat Rev Mol Cell Biol 8:221–233

5. Kandalam V, Basu R, Abraham T et al (2010) Early activation of matrix metalloproteinases underlies the exacerbated systolic and diastolic dysfunction in mice lacking TIMP3 following myocardial infarction. Am J Physiol Heart Circ Physiol 299:H1012–1023

6. Lindsey ML, Zamilpa R (2012) Temporal and spatial expression of matrix metalloproteinases and tissue inhibitors of metalloproteinases following myocardial infarction. Cardiovasc Ther 30:31–41

7. Lindsey ML, Escobar GP, Dobrucki LW et al (2006) Matrix metalloproteinase-9 gene deletion facilitates angiogenesis after myocardial infarction. Am J Physiol Heart Circ Physiol 290:H232–239

8. Lindsey ML, Escobar GP, Mukherjee R et al (2006) Matrix metalloproteinase-7 affects connexin-43 levels, electrical conduction, and survival after myocardial infarction. Circulation 113:2919–2928

9. Agnihotri R, Crawford HC, Haro H et al (2001) Osteopontin, a novel substrate for matrix metalloproteinase-3 (stromelysin-1) and matrix metalloproteinase-7 (matrilysin). J Biol Chem 276:28261–28267

10. Siri A, Knauper V, Veirana N et al (1995) Different susceptibility of small and large human tenascin-C isoforms to degradation by matrix metalloproteinases. J Biol Chem 270:8650–8654

11. von Bredow DC, Nagle RB, Bowden GT et al (1995) Degradation of fibronectin fibrils by matrilysin and characterization of the degradation products. Exp Cell Res 221:83–91

12. Imai K, Yokohama Y, Nakanishi I et al (1995) Matrix metalloproteinase 7 (matrilysin) from human rectal carcinoma cells. Activation of the precursor, interaction with other matrix metalloproteinases and enzymic properties. J Biol Chem 270:6691–6697

13. Lee KH, Choi EY, Hyun MS et al (2007) Association of extracellular cleavage of E-cadherin mediated by MMP-7 with

HGF-induced in vitro invasion in human stomach cancer cells. Eur Surg Res 39:208–215

14. Belaaouaj AA, Li A, Wun TC et al (2000) Matrix metalloproteinases cleave tissue factor pathway inhibitor. Effects on coagulation. J Biol Chem 275:27123–27128

15. Crabbe T, Smith B, O'Connell J et al (1994) Human progelatinase A can be activated by matrilysin. FEBS Lett 345:14–16

16. Trocme C, Gaudin P, Berthier S et al (1998) Human B lymphocytes synthesize the 92-kDa gelatinase, matrix metalloproteinase-9. J Biol Chem 273:20677–20684

17. Birkedal-Hansen H (1995) Proteolytic remodeling of extracellular matrix. Curr Opin Cell Biol 7:728–735

18. Visse R, Nagase H (2003) Matrix metalloproteinases and tissue inhibitors of metalloproteinases: structure, function, and biochemistry. Circ Res 92:827–839

19. Tan RJ, Liu Y (2012) Matrix metalloproteinases in kidney homeostasis and diseases. Am J Physiol Ren Physiol 302:F1351–1361

20. Yu Q, Stamenkovic I (2000) Cell surface-localized matrix metalloproteinase-9 proteolytically activates TGF-beta and promotes tumor invasion and angiogenesis. Genes Dev 14:163–176

21. McCawley LJ, Matrisian LM (2001) Matrix metalloproteinases: they're not just for matrix anymore! Curr Opin Cell Biol 13:534–540

22. Zamilpa R, Lopez EF, Chiao YA et al (2010) Proteomic analysis identifies in vivo candidate matrix metalloproteinase-9 substrates in the left ventricle post-myocardial infarction. Proteomics 10:2214–2223

23. Chiao YA, Zamilpa R, Lopez EF et al (2010) In vivo matrix metalloproteinase-7 substrates identified in the left ventricle post-myocardial infarction using proteomics. J Proteome Res 9:2649–2657

24. Haro H, Crawford HC, Fingleton B et al (2000) Matrix metalloproteinase-7-dependent release of tumor necrosis factor-alpha in a model of herniated disc resorption. J Clin Invest 105:143–150

25. Acuff HB, Carter KJ, Fingleton B et al (2006) Matrix metalloproteinase-9 from bone marrow-derived cells contributes to survival but not growth of tumor cells in the lung microenvironment. Cancer Res 66:259–266

INDEX

Troy A. Baudino (ed.), *Cell-Cell Interactions: Methods and Protocols*, Methods in Molecular Biology, vol. 1066,
DOI 10.1007/978-1-62703-604-7, © Springer Science+Business Media New York 2013

Made in the USA
Middletown, DE
01 November 2022

13899428R00122